ALL AROUND THE YEAR

ALL AROUND THE YEAR

Michael Morpurgo

Photographs by James Ravilious

Drawings by Robin Ravilious

NEW POEMS BY TED HUGHES

JOHN MURRAY

The photographs by James Ravilious
are reproduced by the kind permission of
Beaford Archive, North Devon

Printed in Great Britain by
Cox & Wyman Ltd, London, Fakenham and Reading

0 7195 3577 8

INTRODUCTION

Parsonage Farm is a 210 acre mixed farm near Iddesleigh in North Devon, a few miles north of Dartmoor. It is a country of green, rolling hills, narrow lanes and high hedges, with lichen-covered oaks, fast-flowing rivers, cob and thatched cottages and tall Devon church towers.

Some 50 years ago John Ward's family moved to Parsonage Farm, and he has been farming on this land since then. Hettie, his wife, acts as housewife, mother and farm secretary. Much of the paper work falls on her shoulders, and it is she who looks after them all. Most of the physical work on the farm is done now by John's two sons, Graham who is 24 and David who is 20. Both were born here and both have worked on the farm since they left school a few years ago.

Like Parsonage, the farms around are family farms, many of them run singlehanded. For this area Parsonage is a sizeable farm, many are less than 125 acres. They are predominantly sheep and dairy farms, the hills being particularly well suited to sheep farming. Most of Parsonage land is used for grazing sheep and cattle, but about 35 acres are drilled with barley each year, and very often a further 10 acres with a root crop, that is swedes, turnips and kale.

It is not possible to include in the diary many of the details of life on the farm. To enable the reader to have a clearer picture of the daily routine of work, here is a timetable of any day at Parsonage. It is important to understand that apart from the activities repeated each day, there is always this underlying routine of work.

7.00
Up and out

7.15
Bring the cows in, milk them, wash down the dairy. Feed the pigs and the poultry. Feeding up and checking of cattle and sheep; inside in the winter, out in the fields in the summer.

9.30
Breakfast

1.00
Lunch – main meal

4.45
Tea

5.00
Clean through the milking machinery, get in the cows, milk them, wash down the dairy. Feed the pigs and poultry. Feed up any other stock and shut up for the night.

9.00
Late snack

Stock and machinery are bought and sold throughout a farming year but at the beginning of this year this is an inventory of the animals on the farm, and that is followed by a list of most of the farm machinery.

40 milking cows, Friesians, Jerseys and Guernseys.

50 assorted beef cattle, heifers and calves. Friesians, Devons, Herefords and various crosses.

225 sheep. Mostly Suffolks and Border Leicesters.

22 hens, Rhode Island Red.

9 ducks, Aylesbury.

50 pigs, all Landrace except 1 Tamworth gilt. They include 5 sows, 1 boar, and several weaners, gilts, slips and fatteners.

5 horses. 2 Haflingers, a Dartmoor mare and 2 colts.

Numerous cats and 1 sheep and cattle dog—mongrel, known as Bounce.

FARM MACHINERY INCLUDES:
3 tractors, a combine corn drill, a hay elevator, a hay turner, a mower, a baler, a buck rake, ploughs and harrows, a roller, a tractor loader, a scraper, two trailers, a cattle box, a link box, a dung spreader, a fertiliser spreader, a cattle crush, sheep clippers, and 1 bail milking parlour for 4 cows abreast.

ACKNOWLEDGEMENTS

I have tried to record faithfully over this last year the life and work on one farm in Devon—Parsonage Farm, Iddesleigh. It was not an average year, and Parsonage is not an average farm. Every year on a farm differs from the one before; indeed no two farms and no two farmers are the same. But on every farm, every year follows a similar cycle as the one season merges into the next, dictating its terms to the farmer.

This past year has been spent working closely with the Ward family at Parsonage. It is a book that should have been written by them, but they are too busy. I am deeply indebted to them for their help, their sensitive advice and unfailing patience at my interminable questioning and probing. It is to them that the book is affectionately dedicated.

To Ted Hughes, to James and Robin Ravilious, I extend my deepest thanks. Through their own experience of living and working in this part of Devon, they have added a depth, colour and perspective that I trust will bring the reader closer to the experience of a farming year.

I would acknowledge also the help and advice of many friends and neighbours, especially the help of my wife Clare, who runs Farms for City Children with me, Anne Brough, Osyth Leeston at John Murray, and Mr. Hindson and Mr. Byrne of Taylor, Hindson and McHugh, Veterinary Surgeons of Hatherleigh.

NEW POEMS BY TED HUGHES

SEPTEMBER
Barley. 13

OCTOBER
The Formal Auctioneer. 30

NOVEMBER
Rain. 44

DECEMBER
A Solstice. 60

JANUARY
New Year Exhilaration. 78

FEBRUARY
Couples Under Cover. 96

MARCH
Birth of Rainbow. 110

APRIL
Happy Calf. 126

MAY
Stealing a Trout on a May Morning. 138

JUNE
Last Load. 154

JULY
Foal. 170

AUGUST
Coming Down Through Somerset. 186

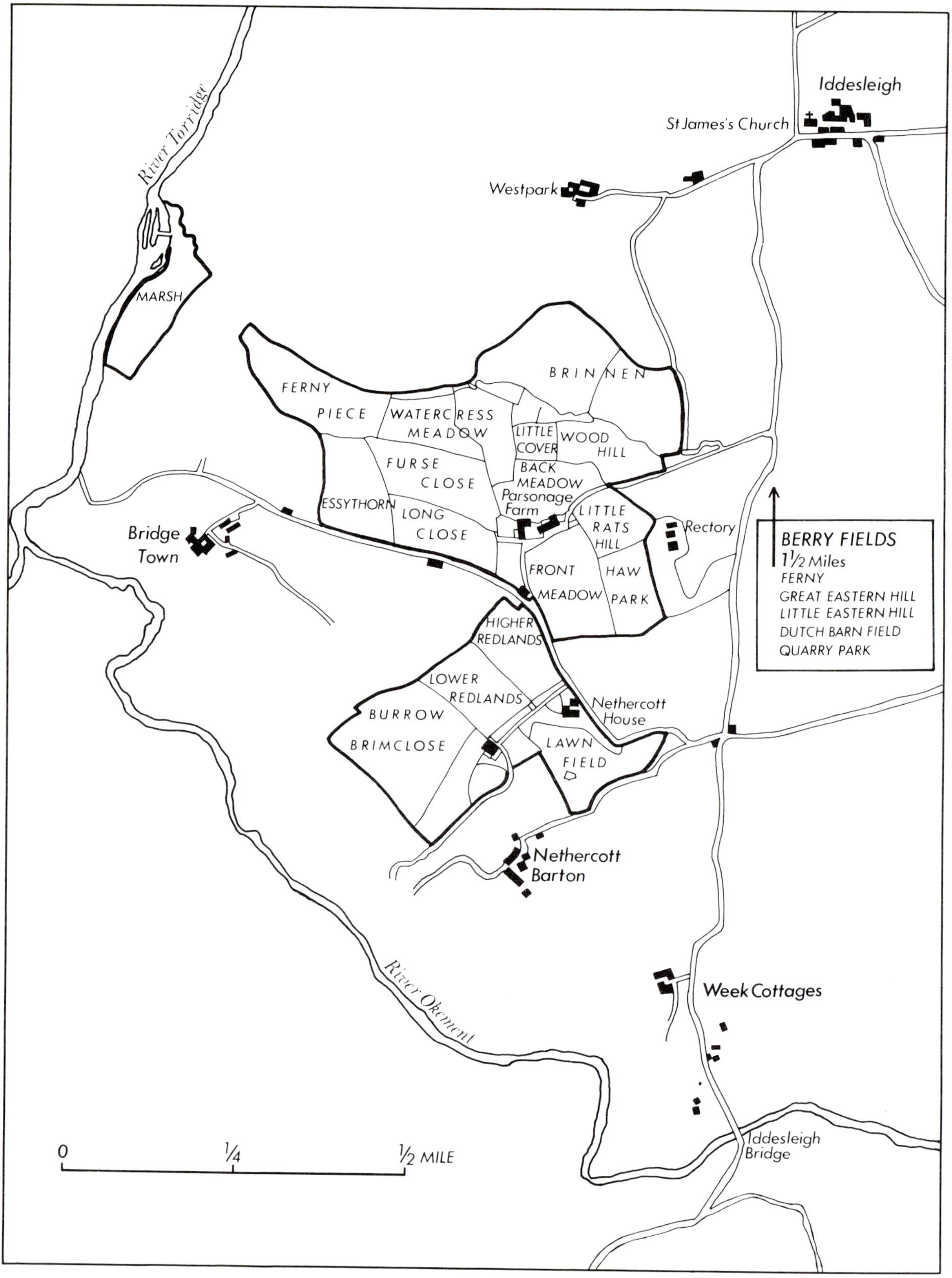
Iddesleigh
St James's Church
Westpark
River Torridge
MARSH
FERNY
PIECE
WATERCRESS
MEADOW
BRINNEN
LITTLE
COVER
WOOD
HILL
FURSE
CLOSE
BACK
MEADOW
Parsonage
Farm
ESSYTHORN
LONG
CLOSE
LITTLE
RATS
HILL
Rectory
Bridge
Town
FRONT
MEADOW
HAW
PARK
BERRY FIELDS
1½ Miles
FERNY
GREAT EASTERN HILL
LITTLE EASTERN HILL
DUTCH BARN FIELD
QUARRY PARK
HIGHER
REDLANDS
LOWER
REDLANDS
BURROW
BRIMCLOSE
Nethercott
House
LAWN
FIELD
Nethercott
Barton
River Okement
Week Cottages
0
¼
½ MILE
Iddesleigh
Bridge

For

John, Hettie, Peter, Graham, David and Elizabeth.

BARLEY

Barley grain is like seeds of gold.
When you turn a heap with a shovel it flows
With the heavy magic of wealth.
Every grain is a sleeping princess—
Her kingdom is still to come.
She sleeps with sealed lips.
Each grain is like a mouth sealed,
Or an eye sealed.
In each mouth the whole bible of barley.
In each eye, the whole sun of barley.
From each single grain, given time,
You could feed the earth.

You treat them rough, dump them into the drill,
Churn them up with a winter supply
Of fertiliser, and steer out onto the tilth
Trailing your wake of grains.

When the field's finished, fresh-damp,
Its stillness is no longer stillness.
The coverlet has been drawn tight again
But now over breathing and dreams.
And water is already bustling to sponge the newcomers.
And the soil, the ancient nurse,
Is assembling everything they will need.
And the angel of earth
Is flying through the field, kissing each one awake.
But it is a hard nursery.
Night and day all through winter huddling naked
They have to listen to the pitiless lessons
Of the freezing constellations
And the rain. If it were not for the sun

Who visits them daily, briefly,
To pray with them, they would lose hope
And give up. With him
They recite the Lord's prayer
And sing a psalm. And sometimes at night
When the moon haunts their field and stares down
Into their beds
They sing a psalm softly together
To keep up their courage.

Once their first leaf shivers, they sing less.
And start working. They cannot miss a day.
They have to get the whole thing right.
Employed by the earth, employed by the sky,
Employed by barley, to be barley.
And now they begin to show their family beauty.
They come charging over the field, under the wind, like
        warriors,
"Terrible as an army with banners"
Barbaric and tireless, a battalion of Amazons.

That is how they win their kingdom.
Then they put on gold, for their coronation.
Each one barbed and feathered, a lithe weapon,
Puts on the crown of her kingdom.
Then the whole fieldful of queens
Swirls in a dance
With their invisible partner, the wind,
Like a single dancer.

That is how barley inherits the kingdom of barley.

Elm trees by the sheep dipping pit

SEPTEMBER

WEDNESDAY 1ST

Tupping. The rams went in with the ewes this morning. We split the flock three ways—one ram to sixty ewes. Only last week John went off to South Molton Sheep Fair and bought two fine young Suffolk rams. We have kept back one of the old ones and he has gone in with the older ewes. Gestation in ewes is 21 weeks so we shall be lambing at the beginning of February.

Our problem with the milk seems to be over. All the summer we have had difficulty keeping the churns cool until the lorry came. Last Thursday we had a letter from the milk depot at Torrington rejecting 30 gallons. We were fairly certain that it was the sun doing the damage but just to be sure Graham and David took the entire milking machine to pieces and sterilised it. Since then there have been no more letters.

TUPPING

The mating between a ram and his ewes. Ewes come in season every 3 weeks so that within that time the ram should have mated with every ewe in the field.

THURSDAY 2ND

Daisy, one of our Guernsey heifers, calved out in the field this morning. She calved down easily and gave us another bull calf—the fifth in a row.

John bought the four Guernseys last autumn at a local farm sale; we wanted the Guernseys to improve the butterfat quality of the milk. The best of them, Celandine, produced a fine heifer calf and now two of the others have had good bull calves. All three of them look promising milkers giving upwards of 4 gallons a day. But the last one has never come in season. We have had her near the house to watch for any signs of bulling, but there has been nothing. She will have to go to market sometime this autumn.

Graham and David have been mending fences all day. We are at the end of a dry summer and there is still no more rain. The brook is not running any more as it was on Sunday, and you can still walk across the Torridge in most places.

FARM SALE

Held at the farm that is being, or has been sold, the stock and machinery of the outgoing farmer is sold off by auction.

FRIDAY 3RD

John has been out ploughing again. The ground is just moist enough now to turn. Last year and the year before we put Little Rat's Hill down to barley, so it is time we gave the soil a rest. This year we will need the extra pasture and the extra

PLOUGHING

Turning the ground over into furrows. The first stage in preparing the ground for drilling.

hay so we will be putting it down to grass this autumn.

It is distinctly colder today and the swallows feel it. They are gathering on the telephone wires down the lane.

One of the ducks died during the night, it is quite common for ducks to get on their backs and then find themselves unable to get up.

SATURDAY 4TH

For some weeks now we have been anxious about the calves in the calf house down at Burrow. Graham and David have been feeding them cows' milk from the bucket, but they have been looking rather poorly. So this morning we have brought them all up into the shippen at Parsonage along with two nurse cows—Petunia, the smallest of the Friesians, and Phyllis 12th. John feels that the calves will do a lot better sucking the milk straight from the nurse cows. This way the milk is drunk slowly and is always at the right temperature. Of course we will go on feeding them their oats and calf nuts and hay.

We are worried about Emily, Emma's calf, born 3 weeks prematurely in May. Emma is one of our two Jerseys and until now has always produced good strong calves, but Emily was only able to survive her first 48 hours with the help of an infra-red lamp. She has never been healthy, and today we had in Mr. Hindson, the vet, to look at her. It seems her heart is weak and she has a lung infection as well. He injected her with an antibiotic, but there is nothing he can do about the heart. He says that after such a premature birth a weak heart is to be expected.

THE VET
Mr. Hindson and Mr. Byrne, work in the local veterinary practice in Hatherleigh.

SUNDAY 5TH

Our milk yield is doing very well, comparatively speaking. There is very little grass in the fields so it is amazing the cows are giving as much milk as they are. Some herds are down by 25 per cent in their yield. Of course we are down as well by half a gallon a day per cow, but considering the dry season we feel quite pleased. To compensate Graham and David started feeding them hay and extra cake weeks ago before they had time to lose condition.

Another problem caused by the lack of grass is magnesium deficiency. There have been several reports of cows dying because of this, so we have put out magnesium blocks on Higher Redlands for them to lick.

MONDAY 6TH

There is no doubt the grass looks greener. We have been having a heavy dew every morning and there was a day or two of heavy rain last week. There should be enough to keep

HEIFER
Female cattle until first calving.

Offloading the milk churns

the sheep going till Christmas if we have just average rainfall.

Graham has been picking out the ewe lambs for tomorrow's market. About thirty of them look good enough to go. He will keep back twenty of the best to go into the flock for breeding.

Jeanie, our best Friesian heifer, calved this evening—a bull calf. First she refused to get on with it out in the field, and it was getting dark so we had to bring her in. Then she would not lie down and seemed more interested in her hay and cake than in calving. We waited and watched hoping she would calve by herself, but she showed no sign of doing so. John decided we had waited long enough and that she was ready to calve. He reached inside and found two front hooves and with Graham helping he managed to tie a rope round each hoof. Then with every push the heifer made they pulled.

It was 40 minutes before the head was out and then the rest slid out quite easily. She seemed unimpressed by her new calf and paid him no attention, refusing even to lick him over. In the end John brought the calf to her and put him under her nose. Only then did she seem to understand what it had all been about. It was a pity she had not grasped it earlier.

Graham checking the sheep

TUESDAY 7TH

We had mixed success at the market in Hatherleigh. Prices for the ewe lambs were a bit disappointing, a lot down on last week's. We had to bring sixteen of them back. Maybe things will have picked up in a few weeks and we will take them back again. But we did very well with the old ewes we took, much better than expected.

Yesterday's calf is doing well, although he is still staggering about on tiptoe. Some calves seem to find their front legs a little stiff after birth and their hooves are not properly formed. It is quite normal, and the calf takes 3 or 4 days until his legs loosen and his hooves settle into the right shape so

that he can put the flat of his hoof on the ground.

Hedge trimming started this morning. It is late because the hedges have not grown out in this drought. We have to call in a contractor for this, Mr. Holwill from Iddesleigh. He has made a start on Little Rat's Hill and Haw Park where John has been ploughing. As usual John has left the headlands so the hedge parings can be picked up cleanly. Once that is done he can go in and plough up the headlands tight to the hedge.

David went into South Molton this afternoon to fetch the grass seed. We sow 30 lb. of seed to the acre and we will be sowing 18 acres of grass—Little Rat's Hill and Haw Park.

HEADLANDS

The headlands are the edges of the field. They are marked out first before the centre of the field is ploughed up. The headlands leave room for the tractor to turn at the end of each furrow. Finally the headlands themselves are ploughed up, usually 7 yards around the perimeter of the field. One year the furrows are turned inwards, the next year outwards. This way the ground is kept level.

WEDNESDAY 8TH

Emily seems a lot happier. She has put on weight and is licking herself—a good sign. But she still spends most of the day curled up and coughing.

David and Graham have been picking up the hedge parings round Little Rat's Hill and Haw Park, and John followed them around the fields and finished off ploughing up the headlands just before dark this evening. Both fields are ready for cultivating now. But rain is forecast.

One of the old sows got out this afternoon and ate her way through a sack of chicken pellets before we caught her at it.

THURSDAY 9TH

The rain came during the night and upset our plans for cultivating today.

The straw arrived this afternoon from up-country, 600 bales of it. We did quite well with our own straw this year, a lot better than last year, but it is going to be a long winter. Certainly there is a month's less grass this autumn because of the drought. We had to start feeding hay and straw to the cattle 6 weeks earlier than usual, and by the end of the winter we shall need every bale of this load. John feels sure there will

Emily and companion

be an increasing demand for hay and straw as the winter comes on, so we should be saving ourselves money by buying it in now. Of course there is every chance that we will have increased the milking herd by the winter months, so our consumption of hay and straw will be up anyway. We had a thin crop—100 bales to the acre—but all excellent quality.

HAY
Grass dried in the field. On this farm we expect on average 120 bales to the acre, though quality and quantity vary from field to field. Hay is our main winter feed for cows, steers, calves, sheep and horses. One bale of hay costs around £1, so it is vital to be as self-sufficient in this as possible.

FRIDAY 10TH
The rain has eased off, but now there's a strong north-westerly wind—too much wind for sowing. However the ground has dried out enough for John to get out and cultivate Little Rat's Hill.

One of the ducks has gone lame. It will have to be killed.

It is the time for fencing and hedging, for ditching and patching up in the buildings. Graham has already made a start on the fences around Watercress Meadow, and this afternoon we were all working on the gates around Parsonage—repairing, rehanging and repainting.

FENCING AND DITCHING
Some farm maintenance goes on all year, but especially in the autumn, a less busy time of year. Ditches are cleared, hedges are cut and patched up. Rotting fencing is replaced, gates rehung and the buildings repaired before the animals are housed in winter.

SATURDAY 11TH
There has been an invasion of sheep down on the Marsh. David discovered them when he checked the lambs this morning. They are a mixed bunch, Suffolks, Dorsets and a few Devon Longwools, but we don't recognise the markings on them. David picked them out and brought them back up to the yard. There is always the chance they might wander back where they came from and take our lambs with them. John has telephoned all the likely farmers, but so far no one has missed them.

SHEEP BREEDS
Suffolk—black faced sheep, very popular, producing lean lambs.

Devon Longwool—resembling a shaggy English sheep dog, a local breed still kept by many farmers. Very good fleece; quiet nature.

Dorset Horn—a stocky, closewool sheep with a round face. Fattens quickly.

Border Leicester—a large leggy, white faced sheep with large Roman nose. Wild in temperament, kept on hills or moorland. Good mothers and milkers.

SUNDAY 12TH
More heavy rain today and there is more on the way. A tinge of fresh green is creeping into the hedgerows and the ditches are running again.

The owner of the wayward sheep has been found. They had wandered over from the other side of the parish. The farmer had not had time to mark them since he had only bought them from the market last week.

MONDAY 13TH
Fat market in Hatherleigh. Graham took in six fat bullocks, a sow and some fat pigs, and the last of the old ewes. He sold them all and did quite well, but prices have gone back since last week. Still, it means less stock to feed through the winter.

John has been to a farm sale over at Morchard Bishop, and he brought back two good looking Friesian cows. He is trying to build up the milking herd gradually this autumn. It is part of a change-over from beef to milk. We need to increase to

Chewing the cud

MILKING PARLOUR
The milking bail is a transportable machine for cows standing abreast. The cows are chained in stalls and milked by machine. The herringbone parlour is more automated. Eight cows (the number depends on the size of the parlour), four-a-side, stand either side of a pit where the dairyman can put the clusters on without bending. Feeding is automatic and the milk is pumped up directly into the bulk tank in the dairy.

LACTATION
The period during which the cow gives milk. Yield reaches a peak soon after calving and is maintained for some months before falling away. A lactation lasts about 305 days.

forty-five or fifty cows to make the fullest use of the land and the new milking parlour we will be putting in at Burrow. At present there is only an old milking bail but we will be replacing it with a herringbone system this winter.

TUESDAY 14TH
Judy and Dolly, the two new cows, seem to be settling well into the herd. There were the usual confrontations in the collecting yard outside the dairy, but the joint leaders, Hyacinth and Angelina, soon sorted the newcomers out. They milked quietly enough, just about 2 gallons each—both are nearing the end of their lactation. Both of them will be calving soon after Christmas just when we will be needing fresh calvers.

Emily is not well, her cough seems worse.

WEDNESDAY 15TH
The nights are crisper now, the evenings drawing in and the young stock need more attention. There is some goodness in the grass, but they need more than that. This morning we

began feeding the lambs some oats and nuts for the first time, and this evening we brought the yearlings inside for warmth and fed them some calf nuts and hay.

THURSDAY 16TH

A fine fresh autumn day and we spent the morning on Berry Hill picking up the hedge parings. Last year we had to rake them up and load them on the trailer, but today we used the new buck rake.

In the afternoon we got on with a job we should have done months before. Years ago a young beech tree blew down in a gale and fell into the brook just below the bridge. Gradually it formed a dam and the brook had to find a way round through the field. John and David took down the big tractor and a heavy chain, and after an hour of pulling and chopping we managed to winch it clear of the banks and haul it away.

BUCK RAKE

Linked hydraulically to the rear of the tractor, shaped like a many-pronged fork, it scoops up the parings. The tractor is reversed along the hedgerow piling up the parings on the prongs.

FRIDAY 17TH

John sharpened up the circular saw and had just started sawing the first load of logs for the winter months when the rain came.

It is still too wct for thc grass seed to go in, and there was a sharp frost last night. We shall have to wait till the spring now.

I think our white cockerel is dying. For 2 days now he has been sitting on his perch, and this afternoon he was found lying underneath. He seems quite unable to move anything but his head. If he's the same tomorrow, he will have to be killed.

CIRCULAR SAW

Set on a cast iron stand the saw is driven by a large belt, powered by the tractor. The speed it turns at is controlled by the throttle on the tractor. Great care is needed, but there is no more efficient way of sawing logs. The noise is deafening.

SATURDAY 18TH

Mr. Hindson, the vet, came this morning to see Emily. Her lungs are in poor shape and she's weak. The problem is that with a premature animal the lungs do not stretch as much as they should in the first few hours of life. As a result they are not as elastic as they ought to be. The lung tissue remains hard and breathing is difficult. However he says that there is a chance that if we can get her through the winter successfully she might pick up when she goes out to grass next spring and has the sun on her back. On his recommendation we have isolated her from the other calves, both to prevent the spread of infection from her coughing and also so that we can look after her more closely.

SUNDAY 19TH

Herma went on to cake this evening. She's due to calve in a fortnight and needs the extra nourishment to build up her milk and to strengthen her and her unborn calf. We call it

CAKE
The term used to describe cattle nuts that contain cereals with added protein and minerals providing extra nourishment. Milking cows are given cake in amounts commensurate with their milk production at the time.

DRY COWS
By feeding less cake and by milking less frequently, the cow is dried off slowly two months before calving.

"steaming up". Herma has been dry for a month now. She was dried off gradually after her lactation, and we are now building her up for her next calving. We would like a heifer calf from her, just so long as her daughter does not inherit her temperament in the dairy. It was Herma that kicked out one of David's front teeth last spring.

MONDAY 20TH

Emily has settled well. Graham made her suck the milk off his fingers and then tickled the top of her mouth as he brought her head downwards into the bucket. She drank it as if her life depended on it. She is not eating her oats and cake

Iddesleigh in high, dry summer

yet, but she will come to it. Her eyes are bright again but the cough is still there.

We've bought ten Maran point-of-lay pullets to supplement our twenty Rhode Island hens who are moulting and going off the lay. The new arrivals were not made welcome and they huddled together in a corner of the run unable to get at the pellets and water. This evening we found them roosting on the top of the hen house. We will have to herd them in every night until the bullying stops and the Marans feel bold enough to go inside.

John has been spreading the last few bags of fertiliser on Burrow Brimclose.

FERTILISER

In granular form, containing nitrogen, phosphate and potash. All grass fields are spread once a year to maintain healthy growth.

POINT-OF-LAY PULLETS

Bought at 5 months old, a point-of-lay pullet should start laying within a few weeks. One sign that a pullet is ready to lay is a large red comb on her head.

TUESDAY 21ST

Mixed fortune at the store market today. Jeanie's bull calf made £62, a good price, but Graham had some trouble selling the twenty weaners he took in and he had to bring some of them back. He was offered £24 each for the larger ones, but he wanted £25, so we will fatten them ourselves and sell later.

WEANERS
Young pigs at 8 weeks old.

STORE MARKET
Animals are bought at the store market, to be reared.

John went off again looking for dairy cows at a good local sale, but the prices were too high—there were cows selling at £600 and more. He came back empty handed.

Graham and David spent the afternoon ditching along Watercress. There is more than 300 yards to do and it is in a bad state, so it will be a long job. Tupping goes on; we changed over the rams today. We have to change them around because it is just possible that one of the new rams is infertile, or that a ewe has not conceived. We have to keep the rams apart. If we were to put them all in together they would fight like stags at rut and very likely kill each other.

WEDNESDAY 22ND

Graham has been at the pigs' house this morning, washing down the walls and cleaning up. The sows will be farrowing again in a month or two, so the cleaner it is the better.

The grass down on the Marsh is looking well. We had the bullocks in there before and they tend to leave behind great wads of grass. Now the lambs have gone in and pulled it about, giving a chance for the fresh grass to come through. Sheep are good for grazing land, providing there are not too many of them and provided they are not left too long in one place. There is an old saying that sheep should not spend two Sundays in one field.

GRASS
Grazing land does not look after itself. To maintain and improve the grass it must be fertilised regularly. It should not be over-stocked nor under-stocked. It should be rested during the winter months to allow early growth in the spring. Good drainage and weed-cutting are vital.

THURSDAY 23RD

John is worried about one of our three Devon steers. He bought them as yearlings last autumn. He noticed last night that one of them was not looking well, so this morning he went back to check again. The steer is limping noticeably and John thinks there could be a split in his near fore-hoof. So this evening we have brought him indoors where we can watch him closely.

FRIDAY 24TH

The greatest weakness in sheep lies in their feet, and Foot-rot and Strip are the commonest problems. Yesterday on his rounds Graham noticed a few of the ewes limping up at Berry, and again this morning we saw some hobbling in Furze Close.

STRIP
An inflammation of the fleshy part of the foot between the cloven hoof. Treatment: scrape away the dirt, trim the hoof with a sharp knife and rub in copper sulphate crystals.

Bounce, the sheepdog puppy, is only learning his trade,

but he did well enough today rounding them up and driving them into the yard. Graham and David managed to catch and treat over twenty of them for Strip. Of course it is a recurring problem when the ground is wet like this, and it is something we have to watch always.

FOOT-ROT
A condition affecting the feet of sheep. More serious but less common than Strip, the rot gets in behind the horn of the hoof. Treatment is the same as for Strip.

SATURDAY 25TH
Herma has calved. John mentioned at breakfast that she was looking like it, but no one paid much attention because we thought it would be a week more at least. Even John was surprised by the speed of it. He went out again later and she had finished the whole operation. Of course it is a bull calf. He is on the small side, but well built. Herma is behaving true to form and won't let anyone near.

There were terrific thunderstorms during the night, and the ditches and brooks are overflowing everywhere. We have had a wonderful crop of mushrooms and blackberries this year. The heavy rain will ruin them.

It has been a sheep day. Down on the Marsh the lambs have been breaking through the hedge and up into Mr. Chamings's fields. We shall have to steep that hedge properly, but first we shall have to move the lambs. We were planning to drench them anyway and then move them on to a clean pasture. Lambs, or hogs as they are called at a year old, need drenching regularly against fluke and worm, and then they need worm-free grazing. Great Eastern Hill up at Berry has had no sheep on it for months, so after we drenched them we moved them up there, driving them through the village.

STEEPING
Hedge-laying. To form a solid hedge, the branches have to be partly cut and then folded down binding the hedge together.

LAMB DRENCHING
Lambs are drenched regularly against the fluke and worm they pick up from the pastures. The drench, or medicine, is squirted down their throats by means of a drench gun. The lambs are then marked to ensure that no lamb is drenched twice. Drenching should aid growth.

SUNDAY 26TH
Graham has been working on the tractors. It is a regular Sunday morning task to grease up and oil them. They need regular servicing and checking, and Sunday seems the best time to do it.

Herma was back down in the dairy this evening. With a cow freshly calved we are not allowed to send away the milk for the first $3\frac{1}{2}$ days. It contains too much colostrum.

TRACTORS
Needed for almost every operation on the farm, ploughing, spreading, tilling, drilling, scraping, raking, haymaking, for pulling the trailers and the cattle box. Expensive but essential to maintain in good working order—a new tractor costs from £4,000 upwards now. A tractor's age is judged not by miles completed but by hours the engine has been running.

COLOSTRUM
The yellow, rich milk the new-born calf needs for a healthy start to life. The dairy cow nearly always produces too much of it for a single calf and therefore the rest goes to any other calves and the pigs. It is not acceptable in the milk depot for human consumption. Colostrum is rich in protein and contains the antibodies that are essential for the calf's survival. A calf should never be deprived of its colostrum.

MONDAY 27TH
Honey was bulling today. Graham saw her being surrounded and mounted out on Back Meadow. Cows come in season every 3 weeks, but we like to leave at least 60 days between calving and insemination. It can be done earlier of course, but a cow should have a rest. An exhausted cow will lose condition, give less milk and very likely her milking life will be shortened. So we shall wait until the next time she's bulling before we inseminate Honey.

Herma and calf

TUESDAY 28TH

We picked up our first windfall apples today. There is a small orchard beyond the pigs' house, Bramleys and cider apples mostly. They will not be ready for picking for a fortnight or so, but in the meantime there is bound to be a mass of windfalls which we send to the cider factory at Winkleigh. They give us about £35 a ton.

Herma turned nasty again this afternoon. Graham was trying to drive Herma and her calf up to join the others. She charged him, head down, and chased him out of the field. Graham decided it might be wiser to leave her where she was. She joined the herd of her own accord half an hour later with Graham looking on.

There was a first class farm sale at Ashreigney and John went. The retiring farmer is known to be a careful and meticulous man. Such sales are rare enough, and everyone was there. Prices went through the roof. John was second bidder on eight of the cows but they went just too high. He has very clear notions of exactly how much he is prepared to pay, and he will go no further.

WEDNESDAY 29TH

David scraped down the cow yard at Burrow and the Big

Barn at Parsonage while Graham scooped out the yearlings' house and the pigs again. We loaded the dung spreader but it is too wet to spread so we shall try to do that tomorrow. Week-in week-out from now on this is a job that has to be done until the stock moves out again in the spring.

THURSDAY 30TH
David has been down on the Marsh making a start on that hedge where the lambs broke through into Mr. Chamings's field. It has not been done for years so it will be a long job. Today he cut away some of the old and rotten wood and hauled it out. He has laid fresh turf up the bank to hold it firm, but the laying of the hedge itself will have to wait.

Rain again, and the milk yield is slipping all the time. Cows do not like continuous driving rain and cold winds, and their discomfort is reflected in the milk yields.

Emily has taken to skipping round her pen. That must be a good sign.

Cows inseminated this month: Poppy, Dahlia, Bonny and Hyacinth II.

DUNG SPREADER
Pulled by the tractor, the dung spreader throws out farmyard manure by means of chains attached to a revolving bar, rotated by power take-off from the tractor.

HEDGING
The banks are worn away by continuous rubbing both by cattle and sheep. The sheep in particular make their own way through, rubbing the bank away. The banks have to be built up by hand or by rear-end loader and turfs laid.

THE FORMAL AUCTIONEER

Is trying to sell cattle. He is like a man
Walking noisily through a copse
Where nothing will be flushed. All eyes watch.
The weathered, rooty, bushy pile of faces,
A snaggle of faces
Like pulled-out and heaped-up old moots,
The natural root archives
Of mid-Devon's mud-lane annals,
Watch and hide inside themselves
Absorbing the figures like weather,
Or if they bid, bid invisibly, visit
The bidding like night-foxes,
Slink in and out of bidding
As if they were no such fools
To be caught interested in anything.
Escaping a bidding with the secret
Celebration of a bargain, a straight gain
And that much now in hand.

When you were among them
Hidden in your own bidding, you stood tall,
A tree with two knot-eyes, immovable,
A root among roots, without leaf,
Buying a bullock, with the eye-gesture
Of a poker-player
Dead-panning his hand. Deep-root weathering
The heat-wave of a bargain.

OCTOBER

FRIDAY 1ST

Today dairy farming went metric—officially. From now on it will be litres instead of gallons. It will take some getting used to.

In spite of the wet, Graham almost finished the ditching along Watercress Meadow. It has been 300 yards of heavy work, but once you start a ditch you must finish it.

Caroline has mastitis. Graham noticed it this morning during testing before milking. As he stripped the cow by hand before putting on the clusters he noticed white globules in the milk. Caroline has it in only one quarter, and that quarter is clearly swollen and hard. Graham injected her after milking.

It is also a blow to the milk yield because for seven milkings after treatment by antibiotics we are not allowed to send milk away. In Caroline's case that's a waste of nearly 20 gallons.

METRIC
Much of farming has been officially metric for some time now. Bullocks are weighed in kilos in market for example. But it is still more usual to talk in hundredweights (cattle), and scores (pigs).

MASTITIS
An infection of the udder that usually manifests itself in one quarter. The quarter may become hardened, and the milk will contain small clots. Wet and cold weather can bring it on, and cows are particularly susceptible when "drying off" during the 8 weeks before calving. It can be treated with an injection of antibiotic up the channel of the teat that is infected. For 3½ days the milk has to be thrown away as it contains antibiotic.

STRIPPING
Milking by hand before the cluster goes on to check there are no mastitis clots in the milk.

SATURDAY 2ND

The milk yield is down again—a direct result of the continuous rain. It is far too early to bring the cows inside off the pastures. We shall just have to hope for a spell of dry weather.

Emily is putting on weight and looking after herself well. Each day she eats her way through her concentrates, and an entire rack of hay. It is only when she coughs that there seems to be anything wrong with her.

SUNDAY 3RD

Graham and David stayed behind for the milking this evening, but the rest of us went off for the Harvest Festival service in the village. The church was packed as it always is for this occasion.

CONCENTRATES
Cattle cake, nuts, pellets—food containing cereal, protein and minerals and fed to stock to provide a balanced diet.

MONDAY 4TH

The recent heavy rain has brought the apples down in their hundreds and we spent most of the day bent double and rummaging around in the long, wet grass for windfalls. We left just the Bramleys on the trees and picked the rest. The tally at the end of the morning was ten sackfuls and Graham

Picking up windfalls

DAIRY
The room adjoining the milking parlour where the churns are kept and the washing equipment. This will house the bulk tank in the future.

HAFLINGER
A breed of Austrian mountain horse, stocky, light brown with white mane and tail. Renowned for their calm temperament and hardy constitution.

will take them into the cider factory sometime this week.

Down in the milking parlour Herma is a different cow altogether. Maternity seems to have mellowed her and now she comes meekly into her stall at number eight or nine, never kicks the clusters off, doesn't shy away at the cold of the hose water. David still keeps his distance. Her calf goes off to market tomorrow, and then she will be alone again.

TUESDAY 5TH
We have been keeping five horses on the farm, a Dartmoor mare and her two colts, one this year's and one from the year before; and a pair of yearling Haflingers, one filly and one gelding. It was horse fair day in Hatherleigh market, and because of the light hay crop we felt some had to be sold.

John took in the mare and two colts and we all went in to give him a hand. Hatherleigh market was noisier and more crowded than usual. The horses ranged from this year's Dartmoor foals, little woolly creatures that are often sold for meat, to tall flashy hunters with evil eyes. Prices went from £5 to £500, and by no means everything was sold. The price of hay is too high.

But John did well. His daughter, Elizabeth's, Raindrops

came into the ring unwillingly and the youngest colt kicked up a fuss, but it did not seem to matter because they all made reasonable prices. It is sad to see them go, but at least we know where they've gone. Raindrops was sold to a farmer nearby, and he wants to use her every day to ride round and check his sheep.

So we are left with the two Haflingers, Fred and Hebe. We will be able to start breaking them in next year.

We didn't take Herma's calf in the end. He was scouring this morning, so Graham drenched him, and he'll stay till next week.

SCOUR
White scour. The main cause of illness and death in calves. It is particularly dangerous in the first week of life. The germ lives in the lower bowel of the animal. Correct feeding and well ventilated, warm housing can help to prevent it.

WEDNESDAY 6TH
Indoor work is building up now. All the dry cows and nurse cows come in by night, as do the yearlings. So we are back to the autumn routine of feeding hay indoors at night and straw out in the fields in the morning. More stock inside means more muck to clear and more muck to spread, and that's what we seem to have been doing all day. We filled the muck spreader and Graham took it down to Watercress to spread it just before milking. It is a lot easier spreading muck than loading it.

Another dry day and the milk yield has gone back up over 80 gallons.

Taking the nurse cows in to suckle the calves

THURSDAY 7TH

The chalk arrived today for the cow cubicles. During the winter months the milking cows lie in under the shelter of the cubicles around the cow yard. This protects them from the extremities of the weather and enables them to lie down in the dry.

Long Close has recovered from the dry summer. A month ago the only sign of the root crop of swedes, kale and turnips that John had tilled back in May, was a few limp leaves and a mass of weed. Now it has been transformed into a tall thick green crop. It will be a valuable food for the cows later on.

CHALK

An ideal base for cow cubicles. It packs down well yet remains porous so that the cows lie on a dry bed of chalk covered by clean straw.

FRIDAY 8TH

Horizontal rain all day, and I had to change trousers three times—nothing seemed to keep it out. Graham and David struggled on with the ditching, but in the end the rain drove them inside.

Our regular weekly delivery of animal feeding stuffs arrived today. The load was typical—sow and weaner meal and barley meal for the pigs, oats and calf nuts for the calves, cattle nuts for the herd, and chicken pellets. But it is the cattle cake for the milking herd that is the single biggest item.

The sheep up at Berry managed to break out on to the road. Luckily no damage done.

ANIMAL FEEDING STUFFS

Most concentrates and compound mixtures cannot be kept for too long. They would go stale in the barns. Each weekly load costs about £350 and this increases as winter comes on and there is less grass available. Cows milk best on fresh grass, but even then they need their cake to keep milk yields up. In winter when hay replaces grass they need even more cake to maintain good milk yields.

SATURDAY 9TH

Graham picked out a dozen lame ewes this morning and trimmed up their feet. It is the wet and the warmth that are doing the damage.

There has been a dead beech up by Old Barn for a year or two and at last we've managed to get it down. John sharpened the chain-saw and he and David set to before lunch. They had the ropes on to pull the beech away from the barn as it fell, but even so the tree fell awkwardly missing the old cob barn by only a few feet. It broke up on impact which will help the sawing. There are dry logs there for a month at least, once we can get them cut up.

Herma broke out this afternoon and tried to barge her way into the calf pens to see her calf. They have been separated now for almost a week. It was lucky she didn't tear her udder on the wire as she jumped the hedge.

SUNDAY 10TH

A crisp autumn morning with a nip in the air and the treetops emerging through the mists in the valley. John and Hettie were out picking the Bramleys in the orchard. There are plenty more to come. We'll be picking through into November.

Emily has ringworm. She has it in two or three places. We have to watch ourselves now because it is very catching—a good wash up to the elbows after any contact should keep it at bay.

Poogly, our dark Jersey and the first cow we bought for the herd, gave us all a fright this afternoon. Graham had let the herd up on to Front Meadow because the grass on the Redlands fields was looking a bit low. Just before tea this afternoon he noticed she was swollen on her left side and at once diagnosed Bloat. He drenched her, and she recovered quickly.

RINGWORM
A fungus that grows on animals, especially when they are housed. It makes an unsightly, crusty lump usually on the face.

MONDAY 11TH

Our best milk yield so far. We sent away 85 gallons this morning; so with a gallon for Emily, a gallon for Herma's calf, 2 gallons for the calves still on the bucket, and another gallon for ourselves, that means it's up to 90 gallons. There are two good reasons for this increase: the autumn calvers are now coming into their full flush of milk and there have been several this autumn, and the grass which has come through green and lush since the end of the summer. Until now we haven't seen a good layer of cream, but Hettie has been scalding a gallon or so three times a week.

BLOAT
This can be a killer if it is not treated in time. It happens when cows move on to fresh grass and overeat. The gasses build up inside and death can occur within an hour.

John was busy all day at a dispersal sale near Bideford. It was dark by the time he got back with six more cows, all Friesians, all young and all good. He had to pay £350 on average for each of them, so he hopes they are as good as they look.

TUESDAY, 12TH

David and Graham are very pleased with the six new cows. They came in quietly, and milked easily, giving 10 gallons between them. Two are almost dry. We've christened the fresh cows Violet, Heather, Sunflower, Sorrel, Marigold and Clover.

Herma's calf was sold in the market this morning. He has recovered from his scour and must have been looking well because he made one of the best prices of the morning—£66. He will go on to someone else's farm now and be reared for beef.

This afternoon we started laying the chalk in the cubicles. It was a dusty business but we managed to finish half of the cubicles by milking time.

Graham turned the sows and the boar out into the yard today. Normally they graze in the orchard by day and lie in at night, but it is acorn and apple time, and they are much more likely to break out at this time of the year. If they do they'll wander miles in search of acorns.

Hettie Ward at work

WEDNESDAY 13TH
Another fine, dry day, and we've been picking more of the Bramley apples to store.

Over 100 gallons of milk—ten churns—went away this morning. The new cows are milking well.

We have almost finished laying the chalk in the cubicles, but there will not be quite enough. We will have to put a layer of sea sand on the top.

THURSDAY 14TH
A terrible night of wind and rain, and there were gales all day long. Trees were down all over the farm, and David had to come away from milking to help saw up a tree that had blocked the lane. And because of the fierce weather David and Graham have been hard at it all day finishing the cubicles.

John decided it was time to bring the bullocks back from the Berry land on the other side of the village. He likes to have all the stock close to the house when there's weather like this. It was an easy drive past the village—they never met a single car and Bounce always seemed to be in the right place at the right time.

FRIDAY 15TH

If we did not know before, we know now which trees and which ditches need attention. The great wooden doors to the cow yard blew off their hinges and it has forced us to replace them with an iron gate. There were hundreds of apples down in the orchard: Graham picked up four sackfuls this afternoon.

Celandine, our first Guernsey to calve, is bulling and we've called the A.I. At the moment we do not want any more Channel Island cows—we have two Jerseys and three Guernseys as it is—so we chose a Charolais bull. A Charolais/Guernsey cross is a fair beef animal.

Bonny was bulling at the same time, and Graham is a little worried about her. She's returned twice—come in season twice after insemination—and she really must take this time. She is our heaviest milker—seven gallons a day as a heifer during her first lactation. But it is very often these heavy milkers that have trouble in conceiving.

SATURDAY 16TH

The rams were changed for the last time. To ensure that every ewe is in lamb we changed the rams round once again. We took the 159 ewes away from the two young rams, and drove them all up through Iddesleigh to Berry to meet the old ram. Then we brought the sixty ewes back from Berry to the two young rams.

Driving sheep up through Iddesleigh has its problems. There are at least twelve different roads and driveways the sheep can take, so we had to keep Bounce working all the time. Sheep must never be hurried as they can overheat and die easily.

Graham's hip is giving him some trouble. It is all the bending down in the parlour and lifting churns. With the new parlour we are putting in none of that will be necessary.

SUNDAY 17TH

Phyllis 1st calved this evening. John had said she looked unsettled, and thought she was going to calve soon. When he went back to do the yard work later on the water bag was already broken.

Phyllis 1st is our oldest cow, about 9 years old and a sixth

A.I.

Artificial insemination. With no bull on the farm, A.I. is used. Run by the Milk Marketing Board this is a system by which farmers can buy the semen of a good bull of the breed they want. A telephone call before 9 in the morning will ensure that the cow in season is served on that same day; 70 per cent of the cows conceive after the first service and 90 per cent after the second. It costs £2.50, and there is no extra charge for one repeat service.

GUERNSEY

A Channel Island breed, famous for its milk. Larger than the Jersey, and smaller than a Friesian, its milk is rich in butterfat. It will give less milk than the Friesian, but more than a Jersey. Golden brown with white patches.

CHAROLAIS

A French breed, large and stocky. Often crossed with home bred cows for beef production. Creamy in colour.

JERSEY

A small Channel Island cow renowned for the quality of its milk which is high in butterfat. Colours vary enormously from light brown to almost black. No white patches like the Guernsey.

FRIESIAN

The most popular breed of cow. Black and white, its qualities for both milk and beef production have made it a familiar sight in the countryside.

HEREFORD

A wonderful beef breed, often crossed with others. Offspring, pure or otherwise, always has a white face.

DEVON

North Devon, a good beef animal, broad and stocky with a dark red/brown coat.

calver. It should have been easy, but moving her away from the other cows must have upset her because she would not settle down in the shippen. Nowhere was right to lie down, and she panicked, charged through the door and out into the night. We managed to drive her back and chain her up. Having gone this far John thought it best to pull the calf off. A bull calf again.

CALVING
An older experienced cow is likely to calve more easily than an heifer, but all calving needs to be watched carefully. From the time the water bag first shows to birth can take between 10 minutes and 2 hours. The calves should stand within half an hour or so and start suckling. The farmer has to decide whether the cow is able to calve down with or without help.

SHIPPEN
A building that houses cattle and calves. Usually an old cob building, the original dairy building on the farm.

MONDAY 18TH
John went off to a large farm sale locally, and bid successfully for two young Friesian cows already in calf to an excellent bull. David went out with the box and brought them home this evening. They look small, but good at around £300 each. Some of the best cows were selling for up to £700. These two will be calving early in the new year, and should be dry during the change-over to the herringbone parlour at the end of December.

TUESDAY 19TH
Heavy rain, so it was difficult to get out to do anything. But David finished off the beech tree and carted away the logs. There is just a stump left now and that will rot down and become part of the hedge. Graham was out tightening up the barbed wire fence on Front Meadow. It's a job for two really, one to pull on the wire and the other to hammer in the staples, but Graham seemed to cope on his own.

All the cows are in again tonight, out of the weather.

WEDNESDAY 20TH
A glorious red glow in the sky when we went down to milking before breakfast, and again this evening when we fetched them in.

The yearlings have developed ringworm now. At least four of them have a crusty growth around their eyes. We must get something done about them.

John wants to nominate his bulls now. It is more expensive, but this way we will know more what we are getting.

NOMINATED BULLS
Unless the farmer specifies, or nominates the bull whose semen he wants, he has simply a choice of breed. The farmer chooses a nominated bull because of its record and looks. However it costs more, between £5 and £10.

THURSDAY 21ST
We milked earlier this morning so that David could get away to the first local foxhunt of the season in the neighbouring parish of Dowland, just beyond the Berry fields. Although the hounds were picking up strong scents no one saw either foxes or cubs.

FRIDAY 22ND
There are two schools of thought on ringworm, the magical and the scientific. It seems we will be trying both. In Hather-

leigh John met the local ringworm expert who specialises in 'blessing' them away. John mentioned the ten yearlings with ringworm and asked him to come up and see them. No need, the man said, he could do it from there. In 3 weeks' time, he said, we would find the ringworm beginning to disappear. Unfortunately John forgot to mention Emily and her ringworm, so the next time the vet's out our way we will ask him to treat her—the scientific way.

One of the sows looks to be close to farrowing. Her udder is low slung and quite pink. It will not be long now, so Graham has moved her back from the yard into the farrowing pen where she will be able to make her nest of straw, and farrow alone.

SATURDAY 23RD

Our milk yield was up over 110 gallons this morning, a record; and our monthly milk cheque came with the morning's post. It came to £950 this month, but we have to take out of that £420 for 4 tons of cake (they eat a ton a week), £250 for hay, £40 for fertiliser and another £25 on water and electricity charges. This leaves a profit of about £100, before tax. Parsonage is a family concern, of course, so we do not take out wages as such. It means really that milking with our small herd can only just break even in the winter months. Of course in the summer when they are out on grass and they need less cake, the cows milk better and the profits are there.

A farmer has to be as good a businessman as he is a good stockman if he is to succeed, and John is continually reviewing each aspect of the farm to make sure it is profitable. The strength of the mixed farm—milk, beef, sheep, pigs and arable—is that the markets are rarely all poor at the same time.

SUNDAY 24TH

The clocks changed this morning and it was light for milking, but this evening it was already dark when we brought the cows in.

We moved Fred and Hebe, the Haflingers, up from Burrow Meadow to Lawn Field where there is still some grass for them. There was nothing left down at Burrow Meadow and they were beginning to poach the land. They are hardy animals and will live out for most of the winter unless the ground gets very wet.

CLOCK CHANGE

Bringing cows in in the dark presents no problem. They know the way. The clock change does not mean shorter working hours. Tractors have headlights and there are lights in all the shippens, and buildings.

MONDAY 25TH

Graham has been rebuilding the bank at the top of Burrow Lane. The lambs ruined that bank last spring when they were beginning to leave their mothers to explore. With the ground now wet through again Graham was able to build it up with earth and stones, laying turf up the side to hold it all in.

Little Eastern Hill up at Berry was sown with barley last year and will be again next year, so John thought the ground could do with a good dressing of farmyard dung. From now on all our farmyard dung will be spread up there. David began with a load this afternoon.

TUESDAY 26TH

The sow farrowed this morning. She had made her nest and was lying down by the time we went in to feed the pigs. She

Rain threatening

farrowed fast, three little pigs coming out in as many minutes, and she finished up with ten young—an average litter. Once free of the umbilical cord the little pigs search for the sow's teats, the strongest ones forcing their way to the teats nearest the head. These are all suckling strongly, although there is one with an enlarged head who looks unlikely to survive.

There are many dangers for the new-born pig. Infection in the sow or in the young can wipe out an entire litter, but deaths occur more often when the sow rolls on the young by accident and crushes them. So Graham always puts the sow inside a crate—a cage—so that the little pigs can feed safely.

PIGS
A sow's gestation period is 16 weeks. A Landrace sow is particularly prolific, producing up to sixteen little pigs at a time. Normally she has eight teats on each side. A litter of sixteen is rare however, and if that number are born, one or two will almost certainly die. Like a cow the sow will not let her milk down to her young unless she wants to. When ready to feed them she lies down on her side.

WEDNESDAY 27TH

The little pig with the enlarged head died in the night, leaving us with nine. The others are all feeding well and the sow seems to be giving them enough. Graham has fixed up a lamp over them to keep them warm during their first few days.

The nurse cows have been changed around. Celandine, our kicking Guernsey, is being brought up to replace Petunia who has gone back to join the herd. Celandine has proved a nuisance in the parlour. She kicks out wildly when the cluster is put on. David and Graham will be glad to be without her. The suckling calves had a rough time this evening. Celandine does not seem to like them any more than she does the clusters.

Graham has started drying off two more cows due to calve in 8 weeks, Lucy and Violet. He milks them once every 2 days, for a week; then every 3 days until they give less than half a gallon. After that the cow is dry and not milked at all until she calves.

CLUSTER

Shaped like an upturned hand with four fingers, the cluster is attached to the cow's four teats by vacuum and sucks rhythmically. The putting on and taking off the cluster can be perhaps the most difficult time in the milking process. The cluster is necessarily of a uniform shape and size, the cow's teats can vary greatly in length and width.

THURSDAY 28TH

The vet came this morning to dehorn the calves. One by one they have to be driven into the cattle crush or cage and their heads trapped between the clamps. Even then Graham and David have to hold the head so as to keep it still enough for the vet. First a local anaesthetic is injected, and then the horn is clipped off and the root burned out.

Phyllis 1st's calf has been scouring and Graham has been treating it with an antibiotic scour powder. The calf was looking better by this evening and should be fit to go to market next week.

DEHORNING

Common practice now in modern farming, to prevent cattle damaging each other and the stockman. Usually done when a calf is under 2 months old when there is only a little horn to remove. If it is not done properly the horn will grow again.

FRIDAY 29TH

Damp, misty morning, and the leaves are coming down fast now. The frost must have helped.

A tractor problem. David took the tractor down to Ferny Piece to load up some sticks, but he forgot to check the diesel gauge. He ran out at the bottom of the hill. Graham is treating Iris for mastitis. She has it in two quarters.

SATURDAY 30TH

Our milk returns (not the cheque) came this morning from the milk factory in Torrington. The returns, which we receive each month, show us the quality of milk we have been sending away, and we've done well. Our milk solids both of butterfat and non-butterfat were well up last month. At the end of 6 months our monthly grades for quality and richness of milk will be averaged out and the bonus is paid on the milk

TRACTOR FUEL

A tractor uses diesel fuel, not petrol, and a supply is kept on the farm. However if the tractor is allowed to run out of diesel and air gets into the system, it has to be blown through and bled until there is no air left. Only then can you refill with diesel.

we supply for the following 6 months. At this rate we shall do very well—our grades are far above average. Of course, there is no doubt that the use of cake is partially responsible, and that is expensive; but both the rich pastures of Front Meadow and the Channel Island cows must have helped.

Graham and David drenched the yearlings this afternoon. Twice a year we drench all the cattle against stomach worm, long worm and liver fluke. We saw the yearlings close to this afternoon during the drenching, and there is no doubt that the ringworm is going.

SUNDAY 31ST

Graham milked alone today—it takes half an hour longer on your own, about 2½ hours from washing through before milking to washing down afterwards. Iris' mastitis is much improved, but we still cannot send away her milk.

Emily is finally weaned off the milk from the bucket. She is much fatter than she was, her appetite for her hay and her concentrates will more than compensate now for the milk; indeed it will sharpen her appetite for the more nutritional foods she needs. Her ringworm makes her look rather bedraggled and sad, but we've been treating her with some powder the vet gave us.

More leaves falling all the time and the ditches are clogged up again already.

The following cows have been inseminated this month:

Tulip—Monday 4th
Buttercup—Thursday 7th
Petunia—Thursday 7th
Beth—Thursday 14th
Patricia—Thursday 14th
Celandine—Friday 15th
Bonny—Friday 15th
Poppy—Sunday 17th
Caroline—Tuesday 19th
Honey—Thursday 21st
Daisy—Thursday 21st
Blossom—Saturday 23rd
Rose—Saturday 23rd
Lily—Wednesday 27th

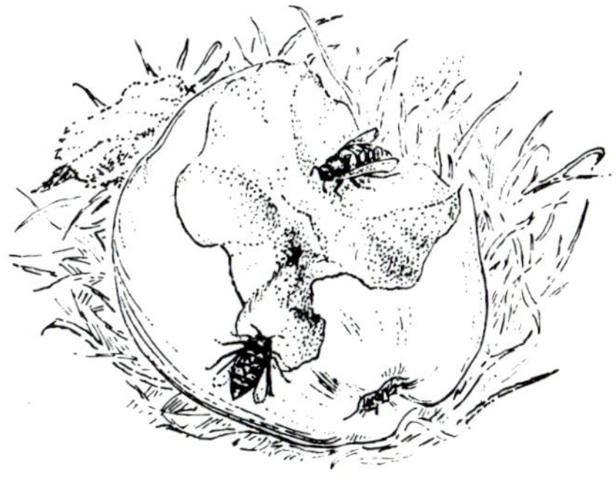

Rain

Rain. Floods. Frost. And after frost, rain.
Dull roof-drumming. Wraith-rain pulsing across
    purple-bare woods
Like light across heaved water. Sleet in it.
And the poor fields, miserable tents of their hedges.
Mist-rain off-world. Hills wallowing
In and out of a grey or silvery dissolution. A farm
    gleaming,
Then all dull in the near drumming. At field-corners
Brown water backing and brimming in grass
Toads hop across rain-hammered roads. Every mutilated
    leaf there
Looks like a frog or a rained-out mouse. Cattle
Wait under blackened backs. We drive post-holes.
They half fill with water before the post goes in.
Mud-water spurts as the iron bar slam-burns
The oak stake-head dry. Cows
Tamed on the waste mudded like a rugby field
Stand and watch, come very close for company
In the rain that goes on and on, and gets colder.
They sniff the wire, sniff the tractor, watch. The hedges
Are straggles of gap. A few haws. Every half-ton cow
Sinks to the fetlock at every sliding stride.
They are ruining their field and they know it.
They look out sideways from under their brows which are
Their only shelter. The sunk scrubby wood

Is a pulverised wreck, rain riddles its holes
To the drowned roots. A pheasant looking black
In his waterproofs, bends at his job in the stubble.
The mid-afternoon dusk soaks into
The soaked thickets. Nothing protects them.
The fox-corpses lie beaten to their bare bones,
Skin beaten off, brains and bowels beaten out.
Nothing but their blueprint bones last in the rain,
Sodden soft. Round their hayracks, calves
Stand in a shine of mud. The gateways
Are deep obstacles of mud. The calves look up, through
plastered forelocks,
Without moving. Nowhere they can go
Is less uncomfortable. The brimming world
And the pouring sky are the only place
For them to be. Fieldfares squeal over, sodden
Toward the sodden wood. A Raven,
Cursing monotonously, goes over fast
And vanishes in rain-mist. Magpies
Shake themselves hopelessly, hop in the spatter. Misery.
Surviving green of ferns and brambles is tumbled
Like a sodden scrapyard. The calves
Wait deep beneath their spines. Cows roar
Then hang their noses to the mud.
Snipe go over, invisible in the dusk,
With their squelching cries.

NOVEMBER

MONDAY 1ST

Fat market in Hatherleigh and Graham took in the two barren cows, Phyllis 12th, and an old Devon cow we used to keep for beef breeding. Phyllis 12th is a bitter disappointment. She is a fine looking cow and should have gone on to produce good calves and a lot of milk—over 6 gallons a day at her best. In the market they made £23 a cwt., and they weighed 12 cwt. (Phyllis) and 11 cwt. (the old red Devon). A fair price but it hardly compensates for the loss. John has had North Devon cows on the farm for as long as he can remember, and was sad to see the last one go. But the accent is now on the milking herd, and no longer on beef, and the North Devon is a beef animal. Of course we shall still keep some bullocks for beef, but we don't have the acreage to build up the milking herd and maintain the same number of beef animals.

TUESDAY 2ND

Phyllis 1st's calf was taken into the store market in Hatherleigh. He was small and only made £45.

It is suddenly a lot colder. Cold water numbs the hands and they stay numb. There is snow on Dartmoor only a few miles away: you can see it from the Redlands fields. We had hailstones, but no snow.

The nine surviving little pigs are looking stronger every day. There is a runt, but he's big enough and strong enough to hold his own. The last time this sow farrowed she had thirteen little pigs and five of them died over a period of a week or so. She did not seem to have the milk to rear them. And her record this time is not good—they are proving a sickly litter. We won't keep her on after this.

WEDNESDAY 3RD

Dairymaid calved by herself during the night. He is the biggest bull calf we've had for a long time, looks a fortnight old already and stands steady and strong. He will be out in the fields with Dairymaid until we sell him in a week or a fortnight. It would be interesting to keep a bull calf, especially when he's as good as this, but it is just not viable financially. If we kept them and reared them we would have to buy in hay and feed them through two winters at £5 a week for every bullock. This way we have a quick return of between £40 to £65 for a week-old calf, and there is no risk of sickness or death.

The last full day of milking in the old milking parlour. Tomorrow we will be knocking down the wall and dragging the old milking bail out on to the concrete by the calves' house. Graham and David won't be sorry. Milking and feeding the thirty or so cows is back-breaking work. With the new herringbone parlour there will be no bending, and it should take half the time. Meanwhile we'll be milking outside.

Graham and I were up at Berry all afternoon, drenching the 150 in-lamb ewes. We had to rig up a pen by the Dutch barn, and Bounce went around and drove them in. The pen has to be as small as possible so that we can catch the sheep easily. Ideally we should have four people on the job: one to catch and hold the sheep, one to administer the drench, one to hold the drench container, and one to mark those that have been drenched so that we don't drench any sheep twice. But, of course, there never are four people available. We managed between us, although holding on to some of the wilder Border Leicesters in the mud was not easy. The secret is to hold them by the neck and pull the head upwards, then they should stop struggling.

Carnival day in Hatherleigh, and the farm had to fit itself round that today. David went off hunting in the morning—there is always a meet in Hatherleigh on Carnival day—and we milked early this evening to be in time for the torchlight procession at 7.30. It was a crisp cold evening, and Hatherleigh was as full as I've ever seen it. There were great flame torches, flaming tar barrels, town bands from all around and unending floats.

THURSDAY 4TH

We are still milking in the old parlour. We made a start and knocked down a wall, but some of the roof had to come away as well. It was a bigger job than we had thought, so we will try to haul it out tomorrow. Milking was draughty this evening.

Emily's ringworm is fast disappearing. She has powders in her feed once a day.

FRIDAY 5TH

The leaves are falling every day now, and the longer perspectives of winter can be seen through the trees again:

Graham took eight of the fat pigs into the abattoir this morning. The live pig market in Hatherleigh has been a bit depressed lately so we thought it better to try the abattoir

Hatherleigh Carnival—the Torch Procession

direct. This way you do not get paid for the live weight of the animal, but have to wait until the animals are slaughtered and weighed dead. They will let us know the dead weight in a few days, and we will be paid so much a score (20 lb.) dead weight.

With Tom Simmonds and Tony Bater, electrician and plumber from the village, we disconnected the bail, took out the glass jars and pipes, hitched it up to the tractor and hauled it out through the cattle yard and round the corner to the calf house. After hours of pushing and pulling and fixing it all back together, it was ready in time for the evening milking. Luckily it was a clear night, but we must put up a temporary roof as soon as we can. The cows came in willingly enough when they heard the rattle of the cake falling into the bins.

SATURDAY 6TH
As expected the milk yield was well down this morning, by almost 10 gallons to 102 gallons. The cows may take a week or more to become accustomed to the new milking situation, so we must expect below average yields. At the same time we have to remember that they have almost finished the last of the grass and what there is cannot have much nutritional value for them. We are increasing their hay and cake all the time to compensate.

The herd joint-leader, Hyacinth, calved this evening at 9 o'clock—a heifer calf at long last.

LEADER
There is a fluctuating hierarchy within a cow herd. There appears to be a leader or group of leaders, but these are often challenged by others. The leader seems to take the initiative, leading them out of fields, and into the parlour. Known as the 'boss cow', it is always interesting to see who takes over the role when she goes dry and is separated from the herd.

SUNDAY 7TH
Hyacinth was lying down when John went in to check her this morning, and she would not get up. No amount of

coaxing would persuade her. It was milk fever, and she had to be treated at once. He called out the vet. Mr. Hindson came within half an hour and gave Hyacinth an injection of calcium. Recovery takes time, so there was nothing to do but wait until the cow stood up, and the waiting was to prove eventful.

MILK FEVER

Milk fever is not uncommon in cows that have just calved. It is a chronic calcium deficiency that will result in death unless something is done to replace the calcium in the blood. Once that is done, then the body mechanism of the cow which is in shock after the birth should take over and circulate its own calcium.

Primrose was due to calve, but she had shown little sign of doing so. While we were waiting in the shed for Hyacinth to get up, she lay down in a corner away from the other cows and started to calve. After a labour of less than 15 minutes she calved—a bull calf. Hyacinth watched the whole operation wide-eyed, then rose majestically to her feet and walked over to the hayrack. So it looks as if all's well.

Both calves are a bit disappointing. Primrose gave us a bull calf that is on the small side, and Hyacinth's heifer is very small with splayed-out forelegs. John thinks it is a fault that is not likely to correct itself.

MONDAY 8TH

Hyacinth is fully recovered and quite back to her old bullying self.

Graham took in thirty-six fat lambs to the fat market. They sold well, the best prices in the market, and that is very satisfying. Six of the best sold for £25 apiece and the others for £23.

But he brought back grim news from the market. Sheep-scab has been reported on a farm no more than 8 miles away.

A cold, showery day, and the cows don't like it. We've begun bringing them inside every night now. Generally speaking it is better for the stock to be outside rather than in, there is less danger of infection; but there is a limit on how long we can leave them out. The cold wet weather does make them lose condition with the consequent loss in milk yield. And, of course, wet fields cannot take continuous grazing, there would be no grass growing up the next spring.

SHEEP-SCAB

All sheep farmers are legally bound to have their sheep dipped against scab at least once a year. Scab is a skin condition that is highly contagious. Caused by parasites, the affected sheep will rub off its wool and lose condition rapidly.

Feeding the sheep

Graham and David had to milk in driving rain this evening. We must get that roof on.

TUESDAY 9TH

David's 21st birthday and Hettie laid on a party in the evening after milking.

David had seen foxes down by the milking parlour before, skulking around in the dark. So tonight he took down his gun and sure enough there they were again, two of them. He sweetened them towards him, much as you would call a cat, and they came slowly until they were in range. David fired and missed.

It is a warning though. He saw those foxes within 100 yards of the chicken run, so we will be more careful than ever now to make sure they are shut up at dusk. All too often recently it has been pitch dark by the time we've got round to shutting them up. And foxes come at dusk and just after.

WEDNESDAY 10TH

Poogly, our original Jersey and the first cow bought for the herd, is being dried off from today. She is due to calve on New Year's Day. A small, dark Jersey, Poogly is unlike other cows and seems to prefer the company of people. However, she milks quite well and has mothered two sound bull calves. She and Emma, the other Jersey, spend most of the time together away from the rest of the herd.

Emily is giving cause for concern again. She has not been eating her concentrates over the last day or so. It is possible she has been having too much concentrated food so we are giving her more hay. She still coughs a lot.

THURSDAY 11TH

The cows seem to have settled now, in spite of the pneumatic drill that has been hammering away in the new dairy for the last 2 days. The milk yield is going back towards the 110 gallon mark again.

Another fine day with dew in the morning and clear blue skies, and that is probably why we still haven't put a roof up over the temporary milking parlour.

We had the returns for the pigs we sold at the abattoir last week. They graded very well, making £36 each. But pig meal is now over £100 a ton, and we get through a ton in a fortnight. The market will improve again no doubt, but at present there is not much profit in pigs.

A lot of scraping and cleaning up today. We took away two loads of dung to spread up on Little Eastern Hill at Berry. Little Eastern Hill is covered with erish, and we will be ploughing in the dung in February sometime for the spring corn.

PIG MEAL
A compound of barley, wheating, minerals and fishmeal.

ERISH
A field of corn stubble.

FRIDAY 12TH

Another heavy dew and mists this morning, but in spite of the sun it is still muddy under foot. There is not enough strength in the autumn sun to dry the land before the next morning's dew.

The stock looks well everywhere, and there is little doubt now that the yearlings' ringworm is disappearing, like Emily's. One cure seems to be as effective as the other.

Graham boiled up the milking equipment this morning. Hygiene in the milking parlour is vital, and we always wash through the milking equipment twice a day after milking, and then scrub down the parlour. Then once a week we give the whole machine a thorough clean through, and remove all the rubber and metal parts to boil them for sterilisation. During these next 6 to 8 weeks in the temporary parlour it is going to be very important to be thorough.

HYGIENE

Even under normal conditions, hygiene is a vital factor in successful milk production. Washing down is an essential part of every milking. Both the parlour and machinery have to be washed thoroughly. If infection and milk contamination is to be avoided, this regular washing down must be supplemented by regular 'spring cleaning' of equipment and buildings.

SATURDAY 13TH

The mists have turned to frost, and all the water pipes to the outdoor parlour were frozen up this morning, so that Graham couldn't even begin milking until 9 o'clock when they had thawed out. The milk lorry comes to the four crossways at 10.45, so time was short. He just made it, arriving at the milkstand with the churns at the same time as the lorry.

When we were feeding hay to the herd after milking, we took away the last bales in the first bay of the Dutch barn. It has disappeared fast and spring is still a long way off. John wakes up at night sometimes and worries whether we shall have enough. At night he never thinks so, but by the morning things seem more promising.

SUNDAY 14TH

Warmer again and we started on the roof over the milking parlour. We put up some wooden supports and picked out some corrugated iron.

Another day of scraping out. David fixed the scraper on to the tractor and scraped down the cattle yard, pushing the dung straight down into the slurry pit which is still only half full.

We selected the bull we want to nominate from the A.I. centre. He's "Snow Regal Olympia", and from now on when we ask the A.I. to come for a cow in the milking herd, he will bring just "Snow Regal Olympia" semen. We have a picture of him, a vast animal with immense forequarters. Let us hope his progeny turn out to be as good as the record shows. It does not always follow.

SLURRY PIT

Conveniently situated just below the cow yard all the muck from the parlour and the cow yard ends up here. With a sloping bottom like a swimming pool, the tractor can scrape the dung straight in. Once it is too full, and if the weather is right, the slurry is loaded into the dung spreader and spread on the fields. A very dangerous place for children, for the surface of the slurry forms into a hard crust, but cracks under pressure like ice on a pond.

David emptying the slurry pit

MONDAY 15TH

The walls of the new parlour are rising already, and the pit has been dug out. If it keeps fine, we should be working in it by Christmas.

Fat market in Hatherleigh. Graham and John picked out three of the bullocks that look ready to go and loaded them up with that Guernsey heifer we couldn't get in calf. The Guernsey made nearly £20 a cwt.; the three bullocks, Devon/Friesian crosses, made almost £30 a cwt. and they were much heavier animals.

TUESDAY 16TH

Graham took Dairymaid's calf into market this morning. He looked a real beauty in the ring and the price reflected this—£74.

The chickens have gone right off the lay now. Normally they should be laying five or six eggs a week each, but the days are shortening and the hens are moulting, so now we are getting only five a day from all thirty-four. They still eat the same amount of food. The Maran pullets we bought, back in September, still haven't started laying. They have grown into huge hens, much bigger than the Rhode Island Reds.

A couple of our Suffolk ewes strayed on to Mr. Yelland's land, and Graham went over this afternoon to drive them back.

HEN BREEDS

Rhode Island Red—a red-brown hen that lays very well. A brown egg layer.

Maran—a black and white speckled hen, not as prolific a layer as the Rhode Island Red. A heavier bird that lays dark brown eggs—but not often.

WEDNESDAY 17TH

We're into the dead season before Christmas, when the days shorten and the farm turns into a mush of mud and leaves.

The hedge trimming was finally completed, and David and Graham will be busy picking up the parings. They have made a start.

The last of the apples were picked up today and Graham piled the sacks into the pick-up and took them into the cider factory at Winkleigh.

Everything seems to be ending and nothing beginning. It is not even real winter yet.

THURSDAY 18TH

Warbling. There are many worms and grubs that cattle pick up from pasture, but about the nastiest is the warble fly. The fly lays its eggs on the legs of the cattle and the grubs eat their way up through the body of a cow until they break through the skin on the back.

So every autumn the entire herd has to be treated. This afternoon we drove the cows into the cattle crush and then Graham poured insecticide along the ridge of the back from the neck to the tail—the area where the warble fly would break through. This permeates the hide and kills the grubs inside. It's a dangerous procedure, and we have to be very careful not to splash any of it on to our skin. It must be done at the right time of year—any later than this and it might kill the animal.

We managed to get the milking herd finished today, it is the turn of the bullocks and yearlings tomorrow.

WARBLING

Every year some cattle die as a result of badly timed warbling. The grub has to be caught at the right stage. Too late and it may already be in the organs of the animal. If untreated the animal would lose condition quickly.

FRIDAY 19TH

We brought the bullocks and yearlings in for warbling. One of the yearlings—Poogly's calf it was—managed to turn

CRUSH
A narrow cage with a gate at each end. Used to confine the cattle for examination and treatment.

CATTLE BEHAVIOUR
Cows are normally docile animals, only likely to be aggressive in defence of young. Bullocks and calves are playful, especially out in the open. When walking through a field where there are bullocks it is always advisable to carry a stick in case they become too playful. Bulls of any breed should never be trusted.

himself round inside the cattle crush and jump the pen gates in his panic. No harm done, but it took several attempts before we managed to persuade him back into the cattle crush.

The cattle are on their worst behaviour at this time of year. They have been out all summer and are not used to being handled. During the deep winter months they become quite tame and lose their fear.

SATURDAY 20TH

One of the bullocks we treated yesterday seemed stiff this morning—affected by the warbling insecticide, but John thinks he will be all right.

A cold morning with all the taps and water troughs frozen up in the valley. The cold excites the horses just as the wind does, and they thundered round and round Lawn Field when I went to see them this morning.

Poogly is finally dried off now, and it is time for her to come up to join the other dry cows at Parsonage, but to separate her from Emma might be asking for trouble so we are leaving her down with the herd for a while. She will have to come away at some stage and the two of them will just have to pine.

Emily seems a lot better now. She is eating everything we give her and she is growing fast. The cough is still there though.

SUNDAY 21ST

Caroline has mastitis again, in a different quarter this time. She is a heavy milker and therefore more prone to it. It's a constant battle to keep down the mastitis level in the herd. We dip the teats in an iodine mixture after each milking and make quite sure we milk the cows right out, and if we are drying them off we always insert antibiotic sticks into the teats to prevent any infection. So it is disappointing every time it is discovered, and particularly in the same cow so soon after treatment—it is only 6 weeks since she last had it. Graham treated her this evening, but it will mean we won't be sending away her milk for 3½ days—seven milkings.

From now on until the spring or anyway while the crop lasts, we shall be feeding kale to the milking herd. This afternoon Graham took the weed-cutter out and cut a stretch of kale off Furze Close, scooped it up on the buck rake and took it down to the cows. We will feed it to them daily now to supplement their dry hay diet.

One of the sows has farrowed this afternoon. It was early and we were caught a little unprepared. Graham saw she was due in a few days and had brought her inside away from the others, but she never looked like farrowing so soon—no one had seen her making her nest. There were twelve in the litter, but they look small and premature and it is not likely they will all last the night.

MONDAY 22ND

Three of the little pigs died during the night, but the others look strong enough to survive now. All three were born very small, one was only half the size he should have been. The survivors are still huddled together under the warmth of the lamp, waking up only to feed.

Graham took four fat bullocks into the fat market in Hatherleigh. They weighed about 12 cwt. each and made nearly £30 a cwt.—over £350 an animal. We shall be selling all the fat bullocks that are ready to go quite soon now. We don't want to have to winter them and the market seems lively at the moment. Graham was pleased—until this afternoon when he drove the tractor into a ditch down at Burrow. He was picking up hedge parings with the buck rake and he went in too close to the ditch. He had to unhitch the buck-rake before he could extricate himself.

TUESDAY 23RD

Primrose's calf was sold in the store market this morning. Small and spindly he made only £41, but we didn't expect any more. Primrose herself is not being very co-operative since she calved. She will let down her milk only to her calf, and that is not convenient in a milking herd. She was the same for 2 months of her last lactation, and then her mood seemed to change and she milked down easily. She has to get used to the milking routine again if she's to be any use, and the sooner the better. She will pine, of course, but that never lasts more than a day or two.

Emily's ringworm has vanished completely now, but one or two of the yearlings still have some around the ears.

SEPARATION

A heifer calf will stay with her mother for 2 or 3 days, mother and calf being brought together twice a day for feeding. A bull calf—and here they are usually sold within a fortnight—is fed by the cow until he is sold.

WEDNESDAY 24TH

John went off to a sale and bought four in-calf heifers, all of them to calve in early January. It was dark by the time he brought them back, so he will be able to see what they look like better in the daylight tomorrow.

THURSDAY 25TH

The full winter feeding for the milking herd has now taken over from the grass. In summer they eat fresh grass and cow cake at milking times. Now in winter it has turned the full circle, and there is no grass left at all.

WINTER FEEDING

Instead of grass, the cows are fed on hay, kale, and sugar beet nuts. The cake ration for each cow is increased substantially during the winter to enable the cows to maintain condition and give down good milk yields. Winter milk yields are always lower than summer. Ideally a cow likes to eat fresh, green grass with the sun on her back.

FRIDAY 26TH

Short wet days with barely enough time to get all the routine work done let alone anything else, but still Graham and David managed to pick up more hedge parings this afternoon.

John and David were exploring the possibilities of opening up an old well this morning. It's on Lower Redlands and years ago it used to be the main supply to the three large meadows on that side of Burrow Lane, but then mains water came and the well became disused. Now with over forty cows drinking up to 14 gallons a day each every day we must use all our available natural water supply. Mains water has become an expensive commodity. John found the well's depth—26 feet, and the water is only 5 feet from the top. He has let in a pipe and is siphoning it away to find out how quickly it refills. If it fills up quickly enough then we'll run a pipe down through to the cattle yard and into the drinking trough.

WELLS

Until 20 years ago, all water on the farm came from the wells, the springs and the streams. There are a dozen wells dotted around the farm, some still in use.

SATURDAY 27TH

The wind is up this evening, and whistling ominously. It will be a wild day tomorrow.

One of the gilts looks as if she might farrow soon.

SUNDAY 28TH

The Landrace gilt farrowed early this morning. She took immense care to make a tidy nest in the straw in the corner of the sty and then produced eight fat little pigs that look strong enough already to be a week old. The old sows have been bullying her and she has been badly gashed on her back recently, but she's showing them the way to bring up young. She is a wonderful mother, and spends all her waking hours snuffling round her litter and grunting her affection at them.

Young sows are often more caring as mothers—it is only when they become older and more experienced that they seem to turn clumsy and trample on the litter.

PIG BREEDS

Landrace—a pink pig of Scandinavian origin. They grow fast and produce large litters. Lean, long carcase. A very popular breed.

Tamworth—a smaller, rusty brown pig. Not so prolific as Landrace.

GILT

Young female pig. A gilt becomes a sow when she farrows for the first time.

MONDAY 29TH

It poured all night and all day. The rivers are up to bank height everywhere. Much more of this and they will be in flood.

Graham took four more of the bullocks into the fat market, and again they sold well, the best of them up over £30 a cwt.

David went ditching on Burrow Brimclose in sheeting rain. It is a vital ditch taking all the water and waste from the milking parlour area down through the meadow. The exist-

BULLOCKS

The term used for older beef cattle. They are killed usually between 18 months and 3 years.

ing ditch was not big enough to take the new increased flow, so it needs not only clearing but widening as well. It's a mammoth task, but it must be done otherwise the meadow will quickly become waterlogged.

The rafters are up on the new parlour and all the milking equipment has arrived. In spite of the cold and the wet, building has been going on apace and we should be working there by the beginning of January. Milking outdoors in January is not a pleasing prospect.

TUESDAY 30TH

Both the Okement and the Torridge are in full flood, great swirling brown gashes in the valleys.

A neighbouring farmer had his sheep stranded by the floods, and John and David spent the morning involved in a rescue operation. The sheep, thirty in-lamb ewes, had got themselves trapped on a high spot in a low-lying field. No amount of cajoling or bullying would shift them, so they had to hitch up the box to the tractor and drive in through the floods. Then they had to set up pens and drive the sheep up into the box.

Mr. Hindson, the vet, was called in to examine Iris, who gave birth last summer to a dead calf, and since then has shown no sign of coming into season. He discovered that she has an infection inside her womb, a sort of cyst. There was a pouch of pus inside her and this was preventing her from coming in season; for the cow there was already a presence in her womb which must have acted like that of an embryo calf. Mr. Hindson pulled out some of the infection himself and then gave her an injection that will induce her to come into season and eject the remaining infection. That will be a false season, but after that she should come back in season after another three weeks.

Modern farming has created its own problems to some extent. In nature there is no good reason why a bovine animal should come in season in the winter months. It is only now when we feed the cows a full nutritional diet all around the year so they can produce milk for us that they come in season at any time. So if from time to time a cow misses a season during the winter months it is hardly surprising.

Again, bulling or coming in season must be helped on by the presence of a bull, and with the use of A.I. that trigger mechanism has often been removed.

Whilst Emma was not looking, Graham took Poogly away from the herd and brought her up with the dry cows. She did not want to leave at all and took every conceivable wrong turning before he finally managed to drive her into Front Meadow with the other dry cows.

A SOLSTICE

Drip-tree stillness. Spring-feeling elation
Of mid-morning oxygen. There is a yeasty simmering
Over the land—all compass points are trembling,
Bristling with starlings, hordes out of Siberia,
Bubbly and hopeful.

We stand in the mist-rawness
Of the sodden earth. Four days to Christmas.
We can hear the grass seeping.
                    Now a wraith-smoke
Writhes up from a far field, condenses
On a frieze of goblin hedge-oaks, sizzling
Like power-pylons in mist.

We ease our way into this landscape.
Casual midnightish draughts, in the soaking stillness.
Itch of starlings is everywhere.
                    The gun
Is old, rust-ugly, single-barrelled, borrowed
For a taste of English sport. And you have come
From eighteen years Australian estrangement
And twelve thousand miles in thin air
To walk again on the small hills of the West,
In the ruby and emerald lights, the leaf-wet oils
Of your memory's masterpiece.
                    Hedge-sparrows
Needle the bramble-mass undergrowth
With their weepy warnings.
                    You have the gun.
We harden our eyes. We are alert.
The gun-muzzle is sniffing. And the broad land
Tautens into wilder, nervier contrasts
Of living and unliving. Our eyes feather over it
As over a touchy detonator.

Bootprints between the ranks of baby barley
Heel and toe we go
Narrowed behind the broad gaze of the gun
Down the long woodside. I am your dog.

Now I get into the wood. I push parallel
And slightly ahead of you—the idea
Is to flush something for the gun's amusement.
I go delicate. I don't want to panic
My listeners into a crouch-freeze.
I want them to keep their initiative
And slip away, confident, impudent,
Out across your front.
                    Pigeons, too far,
Burst up from under the touch
Of our furthest listenings. A bramble
Claws across my knee, and a blackbird
Five yards off explodes its booby-trap
Shattering wetly
Black and yellow alarm-dazzlings, and a long string
Of fireworks down the wood. It settles
To a hacking chatter and that blade-ringing—
Like a flint on an axe-head.
                        I wait.
That startled me too.
I know I am a Gulliver now
Tied by my every slightest move
To a thousand fears. But I move—
And a jay, invisibly somewhere safe,
Starts pretending to tear itself in half
From the mouth backward. With three screams
It scares itself to silence.
                        The whole wood
Has hidden in the wood. Its mossy tunnels
Seem to age as we listen. A raven
Dabs a single charcoal toad-croak
Into the finished picture.

                        I come out
To join you in the field. We need a new plan
To surprise something.
                        But as I come over the wire
You are pointing, silent.
I look. One hundred yards

Down the woodside, somebody
Is watching us.

A strangely dark fox
Motionless in his robe of office
Is watching us. It is a shock.

Too deep in the magic wood, suddenly
We meet the magician.
Then he's away—
A slender figurine, dark and witchy,
A rocking nose-down lollop, and the load of tail
Floating behind him, over the swell of faint corn
Into the long arm of woodland opposite.

The gun does nothing. But we gaze after
Like men who have been given a secret sign.
We are studying the changed expression
Of that straggle of scrub and poor trees
Which is now the disguise of a fox.

And the gun is thinking. The gun
Is working its hunter's magic.
It is transforming us, there in the dull mist,
To two suits of cold armour—
Empty of all but a strange new humming,
A mosquito of primaeval excitements.

And as we start to walk out over the field
The gun smiles.

The fox will be under brambles.
He has set up all his antennae,
His dials are glowing and quivering,
Every hair adjusts itself
To our coming.

Will he wait in the copse
Till we've made our move, as if this were a game
He is interested to play?
Or has he gone through and away over further fields,
Or down and into the blueish mass and secrecy
Of the main wood?

Under a fat oak, where the sparse copse
Joins the main wood, you lean in ambush.

Well out in the field, talking to air
Like quiet cogs, I stroll to the top of the strip—
Then pierce it, clumsy as a bullock, a careless trampling
Like purposeless machinery, towards you,
Noisy enough for you to know
Where not to point your blind gun.

Somewhere between us
The fox is inspecting me, magnified.
And now I tangle all his fears with a silence,
Then a sudden abrupt advance, then again silence,
Then a random change of direction—

And almost immediately—
Almost before I've decided we are serious—
The blast wall hits me, the gun bang bursts
Like a paper bag in my face,
The whole day bursts like a paper bag—

But a new world is created instantly
With no visible change.

I pause. I call. You do not answer.
Everything is just as it had been.
The corroded blackberry leaves,
The crooked naked trees, fingering sky
Are all the usual careful shapes
Of the usual silence.

I go forward. And now I see you,
As if you had missed,
Leaning against your tree, casual.

But between us, on the tussocky ground,
Somebody is struggling with something.
An elegant gentleman, beautifully dressed,
Is struggling there, tangled with something,
And biting at something
With his flashing mouth. It is himself
He is tangled with. I come close
As if I might be of help.
But there is no way out.
It is himself he is biting,
Bending his head far back, and trying
To bite his shoulder. He has no time for me.
Blood beneath him is spoiling

The magnificent sooted russet
Of his overcoat, and the flawless laundering
Of his shirt. He is desperate
To get himself up on his feet,
And if he could catch the broken pain
In his teeth, and pull it out of his shoulder,
He still has some hope, because
The long brown grass is the same
As it was before, and the trees
Have not changed in any way,
And the sky continues the same—

It is doing the impossible deliberately
To set the gun-muzzle at his chest
And funnel that sky-bursting bang
Through a sudden blue pit in his fur
Into the earth beneath him.

He cannot believe it has happened.

His chin sinks forward, and he half-closes his mouth
In a smile
Of ultimate bitterness,
And half closes his eyes
In a fineness beyond pain—

And it is a dead fox in the dank woodland.
And you stand over him
Meeting your first real Ancient Briton
In eighteen years.
And I stand awake—as one wakes
From what feels like a cracking blow on the head.

That second shot has ruined his skin.
We chop his tail off
Thick and long as a forearm, and black.
Then bundle him and his velvet legs
His bag of useless jewels,
The phenomenal technology inside his head,
Into a hole, under a bulldozed stump,
Like picnic rubbish. There the memory ends.

We must have walked away.

DECEMBER

WEDNESDAY 1ST

Graham took the three rams away from the ewes this morning in driving rain. For the rest of the winter the rams will be living out and we shall be feeding them with nuts and oats and flakes to bring them back to condition. By now we hope all the ewes are in-lamb—there are always one or two that will be barren—but if they are not in-lamb by now then we certainly don't want them to conceive after today. Lambs should be born in the cold weather and grow on into the spring and the fresh new grass.

The wet is affecting everything—the hens are not laying, and the milk yield is falling.

THURSDAY 2ND

Sleet with the rain all day. Work has ground to a halt on the new parlour, and on the farm we are reduced to getting done just what is necessary for the stock and nothing else.

FRIDAY 3RD

Dartmoor has turned white this morning against steel grey skies. It is heavy and quiet with snow threatening, but at least it has been dry all day.

In the Big Barn Shamrock is beginning to look as if she might calve soon. She is not due until the 14th, but when John did the yard work this evening she was looking loose at the back and her udder was full enough. It is not quite taut enough yet for the calving to be imminent, but she could calve over the weekend.

John used to do almost everything on the farm, but since David and Graham have been working, he has left a lot of the heavy, physical work to them. At this time of the year John looks after the yard, feeding and bedding the beef cattle, the dry cows, the yearlings and the calves, and driving the nurse cow to the suckling calves. It is 2 hours work in the morning and the same in the evening. Then, of course, there is the management of the farm. Decisions are made jointly, but John is at the helm, using his long experience of the land—some 50 years—to help Graham and David.

David was busy with the weekly sterilising of the milking equipment at Burrow this afternoon, whilst Graham cleaned

BARNS

There are several barns and shippens and calf pens on the farm. Empty in the summer unless used for storage, in the winter they house the stock, the hay and the straw. Some, like Old Barn, have old cob walls, others are more recently built and of a more modern design, big enough to allow the tractor in to scrape down. Ventilation is particularly important for the stock. Calves in particular will lose condition and catch pneumonia in a damp shippen with little ventilation.

out the yard at Parsonage and then picked out two lambs to kill for the deep freeze. We will take them into the butcher next week with more pigs the butcher has asked for.

SATURDAY 4TH

A white, hard frost this morning with mists filling the valleys. At one point the cows were gliding legless over the fields.

The water was frozen up in the temporary parlour, so Graham had to run a pipe from Mr. Rafferty's bathroom at Burrow Cottage all the way across the cow yard to the bail. The cows don't seem to mind the cold as they do the wet. They all lie in under cover of the cubicles and generate their own warmth. Certainly dry frosty weather keeps them cleaner, and makes udder washing before milking easier and faster.

David took over some hay and the hay racks to the ewes on Furze Close. It's the first time they've been fed hay this winter and from now on it will be every day until the spring—two or three bales a day.

Milk recording this evening. The yield of each cow is checked regularly so that feeding adjustments can be made.

SHEEP FEEDING IN WINTER

The sheep stay out in the fields until and during lambing, unless the weather turns very cold. They pick at the grass but there is little nourishment left in it, so they are fed hay to compensate. Breeding ewes have concentrates and oats for about a month before lambing.

MILK RECORDING

The milk is sucked down through the clusters into separate 6 gallon jars. The jars are marked with gallon and pound measures so that individual cow's milk yield can be recorded. Recording takes place once a month and it is on the results observed that the cake ration for each cow is altered.

UDDER WASHING

The start of the milking process. Before milking the udder has to be washed thoroughly so that no dirt is sucked into the jars. This is often the most time-consuming part of milking especially when it is wet and muddy outside. Before the cluster is put on, the udder is dried off so that no water gets into the milk in the jars.

SUNDAY 5TH

Another icy day with pipes and taps and water bowls frozen up everywhere. There were a lot of people running around with hot kettles this morning trying to thaw out the drinking bowls for the cattle and out in the fields there was half an inch of ice over the drinking troughs.

Iris came in season as the vet said she would, so we can take it that the cyst is all gone now. Next time she comes in season we will have her inseminated. Dairymaid was bulling this afternoon, but it is too soon for her this time.

It is the Christmas livestock show in Hatherleigh market next week and John is thinking that he might have shown this year if he had thought about it earlier. One of those three North Devon steers he bought last year is looking quite good enough. But for showing you really need to have the animal inside and to feed him up so that he fills out everywhere.

There is a genuine purpose in showing, other than the mere winning of prizes. It is very important to maintain and improve the quality in breeding, and this is one way of encouraging it.

Bounce is missing. He disappeared after milking this evening. Usually he comes up with David and Graham, trotting behind the tractor, and then spends the night in the store shed. But there's no sign of him tonight.

MONDAY 6TH
Bounce has reappeared. David found him curled up in the Dutch barn down at Burrow when he went to milking this morning—clearly it was too much like hard work to walk back home last night.

Graham took in the four fat pigs to the butcher. Everything is off the land now except the sheep.

Angelina has returned again for the fourth time. She has been inseminated three times with no success. John called in Mr. Hindson to have a look at her, but he is not optimistic. If she were to return once again after this his advice was to get rid of her. There is only a fifty-fifty chance of her conceiving now, but she was inseminated again this afternoon. It is her last chance.

If she does have to go it will be a great disappointment. She is a huge Friesian, possibly the biggest in the herd, but a cow that gives no milk is no use in a dairy herd. She's also the joint leader with Hyacinth.

TUESDAY 7TH
Graham castrated a litter of pigs. It was not a pleasant business, but it was over very quickly and after a day or so they won't be bothered by it. He used a surgical knife and was careful to sterilise before and after the cutting.

CASTRATING PIGS
This has to be done for two reasons. Firstly, all breeding has to be extremely selective, and pigs come to maturity by 5 months. Secondly, some say that the meat tastes rather strong in uncastrated pigs, so the butchers do not like them so much. Once castrated, they are called barrow pigs.

WEDNESDAY 8TH
The end is in sight at last along the ditch on Burrow Meadow. They've finished more than two-thirds of it now, but the last stretch looks more awkward than ever, with overhanging branches and tall brambles to cut away. It should be done by Christmas.

John is very pleased with the cattle around the yard. The calves look silky and bright, and seem to thrive on the nurse cow. It is at this time of year he will get to know the cattle well, their appetites, their habits and their vices. Hyacinth's little heifer calf, the one with the splayed out forelegs, has come on well and although there is a hint of the problem still there, John feels sure she will make a fair cow for the milking herd.

THURSDAY 9TH
Shamrock calved this afternoon just before dark. John took her away from the others after he had fed them this morning and put her by herself in the shippen. She calved unassisted late this afternoon—a heifer calf. So now we have five heifer calves under 6 months old—that's five heifers and twenty-five bull calves. This one looks to be about the best of all the heifer calves. She's Shamrock's fourth calf, and she's a good,

high yielding cow, just the type we need to breed heifers from.

Graham has been ditching with David all afternoon. It has been a freezing day and they had to work hard to keep warm. They have been lopping off the overhanging branches, and laying them back so that the sun can get to the grass tight to the bottom of the hedge.

FRIDAY 10TH

There are still eighteen of this year's lambs left to fatten, and we are feeding them up with hay and concentrates. We shan't keep them on through the winter—they will be sold as soon as they are fat enough to grade. These are the smallest and the slowest to fatten of this year's lambs—the vast majority were all sold in the spring and summer. We still have the forty-five best ewe lambs to be kept back for breeding.

GRADING

All fat stock (cattle, sheep, pigs) sold in the market has to reach a minimum standard. There are inspectors from the Ministry on hand to check this. Each animal must satisfy the inspector that it would make a good finished carcass.

Feeding hay to the bullocks

SATURDAY 11TH

It was white frost again this morning, but the sun came through warm by midday, and it thawed for 3 or 4 hours before it froze over again. Working in the cold has its problems. Of course the feet go numb, but you can forget that after a time. It is the hands that seem to suffer most from the continual contact with cold water which causes them to crack and chap.

The cows are eating half a bale of hay a day each now—20 lb. of hay—and that would be three-quarters of a bale without the kale and sugar beet nuts. Already the second bay of the Dutch barn at Burrow is looking empty. We had hoped to have enough there to last the milking herd until mid-January, but it does not look very promising at present.

HAY BARNS

There are four on the farm. One Dutch barn at Berry and another by the milking parlour down at Burrow. However, Big Barn has the greatest capacity. It is here that all the straw is stored and some of the hay. Old Barn is also used for hay storage if necessary.

SUNDAY 12TH

A quiet Sunday until breakfast, when the two Haflingers

were seen trotting up the road to Monkokehampton. They came to hand easily enough and we led them back. Somehow they had demolished the iron gate on Lawn Field. From the skid marks it looked as if they might have slid into it accidentally, broken it and seen their way clear.

Graham picked out two more fat pigs to go to the butcher tomorrow. Pigs are not making money. The pig meal is very expensive and if we started adding up the work that has to go into them, there is little doubt we would see a loss. Sows will be farrowing over the next month so the more fat ones we sell now, the cheaper the feeding costs will be. They are ready, and the butcher wants them, so they will go. We shan't come out of pigs altogether. The great advantage of a mixed farm is that you concentrate on the most profitable, minimise losses on the rest and wait for the markets to change. This time next year the rearing of beef cattle may be unprofitable, or everyone might think lamb was too expensive to eat and those markets would be depressed.

MIXED FARMING

Farming is a risky business with increasingly complex factors affecting the markets. The vagaries of weather and governments make long-term planning difficult on a small farm. The safest, and indeed the traditional way, is to spread the risk by mixed farming. At Parsonage Farm with barley, sheep, a milking herd, beef bullocks and the pigs, it is as certain as anything can be that something will be profitable.

MILK YIELDS

A Friesian should milk an average of 6–7 gallons a day in her full flush of milk. A Guernsey 5 gallons a day, a Jersey 4 gallons a day. A cow will maintain its best level for 10 weeks after calving then it will steadily fall until drying off begins 8 weeks before calving. A Friesian gives about 1,200 gallons in 305 days' lactation. A Guernsey gives 1,000 gallons, and a Jersey gives 800 gallons over the same period.

MONDAY 13TH

Shamrock's calf was scouring a bit this morning, so Graham has drenched her. As always with scouring the sooner it is treated the better. This one is a really promising calf; she looks better every day.

Patricia is rapidly earning herself the reputation of the poorest cow in the herd. At her very best, in the full flush of milk, she was milking only 4 gallons a day. But she has developed another problem as well—she has started to suck her own teats. So that of the little milk she does give, not all goes into the churn. We shall have to find a way of stopping her.

TUESDAY 14TH

Warmer today, and we are back with the mud, and mud when there's been frost on it seems stickier than ever. David took the opportunity to scrape down everywhere, the cow yard, the calves' house, the shippen and the big barn. He loaded up the dung spreader and spread it over on Little Eastern Hill.

Graham has been putting up a fence along the ditch on Burrow Brimclose. The ground is soft enough now and the stakes went in easily.

WEDNESDAY 15TH

Graham ploughed up a corner of Lawn Field this afternoon. We have decided to grow a few rows of potatoes next year, and this is the best place to put them. With potatoes the price they are we want to grow all we need for ourselves and sell the rest.

In the afternoon David took away the forty-five ewe lambs we are keeping for breeding, and drove them down to Marsh where we will be feeding them hay each day now. There is still a fair amount of grass down there, so they won't need much hay just yet. The others we shall be taking up to Berry soon until lambing time in mid-January.

THURSDAY 16TH

John turned his knee awkwardly while he was doing the yard work this morning, and now it is swollen up badly. He had trouble with the same knee many years ago.

Violet is nearly dry now, and she will be coming up to Parsonage away from the main herd soon.

FRIDAY 17TH

John's knee is a lot worse this morning. He managed to get out to do his yard work, but he should not have tried. He went to see the doctor in Hatherleigh who says it is fluid on the knee and he will have to rest it. David and Graham will share out the yard work between them after milking.

Patricia was caught red-handed in the cubicles when David went down to get the herd in for milking. She was lying there sucking her own teats. We've still not managed to get a muzzle for her.

SATURDAY 18TH

Graham has noticed that Honey is off her hay and her kale—she is still eating her cake in the parlour but that's all. He thinks the cause goes back to haymaking last summer. We had spread slurry on the Redlands fields, and because there was never any rain to help it soak in, it dried in lumps which were baled up with the hay when it was cut. Honey must have eaten more than was good for her, and it has upset her.

SICK ANIMALS

In the first instance, unless there are clearly dangerous and obvious symptoms, the farmer does not call in the vet. He has to make his own judgement based on his experience. Vets cost money and are called out only if the condition is too serious to be treated by the farmer.

Christmas is almost on us and Hettie has been preparing for weeks in the kitchen. Her work is an integral part of the working of the farm. She cooks a big breakfast and a big midday meal for the men, and there is a never-ending supply of clean clothes to provide. She is also the family's V.A.T. accountant, and is the unappointed chairman when there are differences of opinion on how the farm should be run.

SUNDAY 19TH

Graham and David drove the ewes up through Iddesleigh to Berry today. They will be there for the last few weeks before lambing.

We had our Christmas carol service in Iddesleigh this evening. Church bells rang out in the darkness, then carols,

David cutting kale

handbells and readings in a full church. There will be more carol singing around the village on Christmas Eve.

MONDAY 20TH

It was Christmas poultry market in Hatherleigh today, and David and Hettie spent most of the day waiting to bid. With hundreds of chickens, capons, turkeys and geese, each one coming under the auctioneer's hammer individually, it was bound to be a long wait. Hettie came away with a 13 lb. goose, a 20 lb. turkey and two large capons. She was particularly pleased with the goose, a good one amongst a poor batch. Geese were selling for a pound a pound or thereabouts, but Hettie bought hers for under 70p a pound.

Graham spent the day cutting kale, picking up hedge parings down at Burrow, and cleaning out the pigpens.

I heard this morning the first bleating of a new-born lamb from over on Mr. Yelland's fields. He has Dorsets and is lambing a lot earlier than us, but it is a sound that heralds the end of the dead months and the beginning of a new year.

LAMBING

Farmers time their lambing as they wish. Some begin in December and finish within a month. Some spread the load over January, February and March. Many begin as spring comes. A few farmers now have two lambing seasons in a 12 month period. It is important, however, for lambs to be born in the cold of winter. Hot sun is harmful to new-born lambs.

TUESDAY 21ST

Honey is causing us some anxiety. She has been losing condition for some days now. She has lost her appetite, her coat is standing out and she hangs her head all the time. Clearly there is something wrong, and something more than indigestion. It has gone on too long for that. The vet was called in to check her over. He thought it could be one of three things—magnesium deficiency, weed poisoning of some kind or liver fluke. She has been treated for all three, and if she has not improved by tomorrow, then we will have to get the vet back.

WEDNESDAY 22ND

Honey is a lot better this morning. She's eating ravenously, and already her coat looks shiny. One of the remedies must have been the right one.

The nine pigs born last October 26th were finally weaned today at 8 weeks old. Graham took away the sow this morning, but he'll bring her back for 2 or 3 days to let them suckle, just to soften the break. Of course, weaning is a gradual process. Over the last few weeks the little pigs have been fed an increasing amount of pig meal, but from now on there will be no more sow's milk to complement the diet. They no longer have any need of it and anyway they will fatten faster on pig meal.

PIGS

When first born a little pig weighs about 2 lb. At 8 weeks he weighs around 40 lb.

THURSDAY 23RD

Dolly, one of the two cows John bought last month, has calved. In the sale catalogue she was due to calve in mid-January, but there must have been a mistake because she was quite ready to calve with a full udder, and the calf is large and well-formed. She had been inseminated by a Hereford bull and from that we would have liked a good Hereford cross Friesian bull calf, but we have a heifer calf. Hereford cross cows are no use in a milking herd; so she will be sold as a 2-week-old calf and she will not make much, about £30, as opposed to £60 for a bull calf.

The hitching link on the yard scraper broke this morning and Graham took it over to Mr. Vanstone, the blacksmith in Monkokehampton, just 2 miles away. He welded it back on in an hour and this afternoon Graham scraped down the cow yard.

With the walls of the new milking parlour ready to paint, David has made a start. The paint protects the wall, and makes cleaning down easier. It will be a clear sky blue colour and already with only half a wall done the whole place looks brighter. He will be spending up to 4 hours every day working there and he is anxious to make it look as pleasant as

BLACKSMITH

A good, local blacksmith is a great help to the farmer. Broken machinery has to be mended as quickly as possible. With older machinery the blacksmith can make up the part needed when parts are no longer available on the market.

he can. He is still down there now. I can see the light from my window, and it is past 11 o'clock.

FRIDAY 24TH

Judy calved by herself during the night. She is a high yielding Friesian, inseminated with a Friesian bull and so naturally we would have preferred a heifer calf to go into the herd in a couple of years. But it was a bull calf. She is a very protective mother and lowers her head if anyone goes in too close, but both are doing well.

Poppy looks as if she is the best cow we have. She calved back in hot July, and was giving nearly 8 gallons a day at her best. Even now she is still giving over 5 a day. John bought her in Hatherleigh market and paid less than £200 for her. She does not look the part, a bit small and scraggy. Angelina on the other hand was twice the price and milked only 5 gallons in her full flush of milk. Now she is down to 2 and we can't get her in calf. If she is not in calf by the end of January she will have to be sold off as a barren cow.

PRICE OF COWS

Ordinary milking cows cost anything between £250 and £500. If they have a recorded pedigree or exceptional milking figures it could be much more. Price depends on pedigree, age, fitness, shape of body, size and shape of udder, the owner-recorded milk production and many other variables. The farmer has simply to back his own judgement. The best looking cow in the sale ring can disappoint when you bring her home.

SATURDAY 25TH

Christmas Day. Church at 8 a.m. was a quiet affair with just a few people, but the bells sounded well in the cold morning air with only the mist between the church and Dartmoor.

All day the temperature never rose above zero. Picturesque certainly, but all the troughs and drinking bowls needed thawing out with hot water.

It is a day when we do no more than we have to, but at this time of year that is quite a lot. We were kept busy till lunch time, milking, feeding the cattle, the pigs and the sheep. We got as much done as we could, knowing from experience that no one would feel much like doing anything after goose and Christmas pudding.

It was too frosty to cut the kale, so the cows needed extra hay in the yard. With Dolly and Judy freshly calved, we are back over 100 gallons again.

SUNDAY 26TH

The freeze looks as if it is here to stay, and it has come as a shock. For some years now we have not really had a hard winter, and we had forgotten the problems it can bring.

One of the few remaining last year's lambs seems to have gone blind. She was lying away from the rest of the flock in Upper Redlands when I approached her and she never moved. I had to touch her before she knew I was there. A pity, but she should be going to market quite soon now anyway.

BLINDNESS

Caused by a knock, or by disease, an animal that goes blind in both eyes is killed unless there is a chance of recovery. Two of the most common diseases that cause blindness are New Forest Disease in cattle, and Toxaemia in sheep.

Dairymaid was bulling this morning, but it has not been

Collecting a feed bin

60 days since she calved, so we shall leave her till the next time she comes in season.

MONDAY 27TH

It is too cold again for kale cutting. Apart from the difficulty in cutting it, it would have to thaw out before we could give it to the cows. Frosty kale can be a killer. The cows don't seem to miss it, but of course they need more hay to compensate and this is playing havoc with our hay stocks.

The hardest frost so far, and it stayed white till nightfall. When the mists cleared this morning I could see the moor was covered in snow. It is in weather like this that you notice the birds more. They seem to be the only things moving in a still landscape, and their colours are clean and clear against the white all around.

There was a phone call from Dowland this evening, just as it was getting dark. Our ewes had got out from the Berry fields and were heading off in the direction of Dolton. Graham rushed off to fetch them back, but it was dark by the time he found them, and it is not easy to drive sheep in the dark.

The next sow to farrow looks as if she might be close. Graham has put her in the farrowing pen by herself.

TUESDAY 28TH

We were not due to bring the sheep back from Berry just yet, but last night's break-out has made it more urgent. So this afternoon we drove them all back through the village. They came back fairly easily except for two wild Leicesters. They decided they would go another way home, and we had to enlist the help of some boys to get them back.

By the middle of the day it did warm up sufficiently for David to cut a trailer load of kale—enough for 2 days if it freezes again. The cows were delighted to have it once more and showed their pleasure by dancing round the trailer when it arrived.

Like ourselves, all the animals need more food to keep them going during the cold. In particular the in-lamb ewes, now only 3 weeks away from lambing, need extra food, so we are increasing their concentrates.

WEDNESDAY 29TH

Poogly is due to calve on New Year's Day, but it doesn't look as if she is going to make it. She is not yet showing a full udder. But she could surprise us.

Several degrees of frost and hot kettles were in great demand all day. Only the water trough in the cattle yard did not freeze over. There is only one trough between thirty-five cows, so they must have been drinking often enough to stop the ice forming.

The sow will farrow soon, no doubt of that. She was making a nest in the straw last thing this evening. It should be tomorrow.

THURSDAY 30TH

The frost and the ice have gone and we are back to normal with the ditches running and mud on our boots. The taps work when we turn them on and our hands no longer burn hot and cold when we touch cold water first thing in the morning.

Before lunch the sow had farrowed, ten little pigs and she took scarcely an hour to do it.

Shortly after lunch a calf was born over in the big barn. Last autumn John bought in four young Friesian heifers from a good milking herd: it was the first of these that calved this afternoon. She had been running with a Hereford bull, so we were not sure of the precise calving date. Anyway she did not need us and calved down easily enough for a first calving heifer. A bull calf would have been better, but we've got a heifer. The mother, Orchid, seems hardly interested in her calf and won't let her suckle for long without kicking her away. She has to be tied up so that the calf can take the milk she needs. Heifers are often like this with their first calf.

CROSS-BREEDING

The purpose of cross-breeding is to marry the characteristics of one breed with another to enhance the commercial value of the offspring. Cross-breeding for milk is not so common, although there are many Friesian/Ayrshire crosses. Beef cattle cross-breeds on the farm are:

Hereford cross Friesian—the most common cross, providing beef from the dairy herd. The best cows kept pure bred to provide the followers for the dairy herd, and others crossed with Hereford for beef.

Devon cross Friesian—good quality beef animal, black in colour.

Charolais cross Guernsey—good secondary beef animal, fawn in colour.

Devon cross Jersey—red with faint black stripes. A small secondary beef animal.

The Christmas party in Iddesleigh tonight, and we all went. The village hall can take about a hundred, and that's the number there were. It was a party for everyone—children to grandparents.

FRIDAY 31ST

The ewes were moved over to Burrow Brimclose this morning. They are having about 2 lb. of concentrates a day each until lambing time. That way they can build up enough milk to feed two or even three lambs if they have to.

One loss last night amongst the new litter of little pigs. The runt was found dead, crushed by the sow. The sow was already in a crate to prevent this, but even then the weaker ones can get caught. A sow weighs two hundred times the weight of a little pig.

RUNT

In many litters there are one or two smaller, weaker little pigs. They are usually born last and have little chance of survival.

The milk yield is well up again, over 110 gallons. This may be because there are cows freshly calved, but I think it may also be a change of hay. This batch we are feeding them now is fresh and sweet-smelling, a lot better.

Another sow is due soon and Graham brought her into the sty and has put her in a crate already. When she farrows, and it won't be long, there will be three litters of pigs in there. It is quieter feeding the pigs now. The older ones that made all the noise have gone to the butcher, and the little ones just sleep and twitch.

Poogly should calve tomorrow, but Poogly never does anything when she should.

The following cows have been inseminated during this month:

Angelina—Monday 6th
Iris—Tuesday 14th
Phyllis—Thursday 29th

NEW YEAR EXHILARATION

Finds its proper weather on the 3rd day. Pressure
Climbing and the hard blue sky
Scoured by gales. The world's being
Swept clean. Twigs that can't cling
Go flying, last leaves ripped off
Bowl along roads like daring mice. Imagine
The new moon hightide sea under this
Rolling of air-weights. Exhilaration
Lashes everything. Windows flash,
White houses dazzle, fields glow red.
Seas pour in over the land, invisible maelstroms
Set the house-joints creaking. Every twig-end
Writes its circles and the earth
Is massaged with roots. The powers of hills
Hold their bright faces in the wind-shine.
The hills are being honed. The river
Thunders like a factory, its weirs
Are tremendous engines. People
Walk precariously, the whole landscape
Is imperilled, like a tarpaulin
With the wind under it. "It nearly
Blew me up the chymbley!" And a laugh
Blows away like a hat.

JANUARY

SATURDAY 1ST

It has been a disappointing New Year's Day. By feeding time this morning the sow had farrowed, but there were only five born and of these two were already dead. Even the three survivors look feeble. It had not been a particularly cold night, and although Graham likes to have a new-born litter under the warmth of a lamp fairly quickly, he does not think it could have been the cold that killed them. It was just a weak litter. As for the sow, she will be going back to the boar to try again.

A sow cannot come into season whilst feeding young, so the three little pigs needed a foster mother. Luckily we had that farrowing last week and that sow is quite capable of rearing three more. A good sow can rear up to sixteen, so thirteen should not be too much of a strain. The problem is rejection. After 3 or 4 days a sow will reject any new pigs added to her own litter. However this adoption appears to have been successful. By this evening she seemed quite happy to let them all suckle and the three outsiders were in there fighting along with the others.

Last year's lambs become 'hoggets' today, teenage sheep if you like.

No sign that our Poogly is to calve. We think we may have got the A.I. date wrong; perhaps we recorded the date she was first inseminated and not the second.

SUNDAY 2ND

Another of the heifers has calved. Pansy, the second of the batch of four heifers John bought in last year, is a fine deep-bodied Friesian and she's calved a stocky Hereford cross bull calf. The calf pens are full of little white faces, but not for long. Some or all of them will be going off to store market on Tuesday in Hatherleigh.

The time has come now for Emily to move back in with the other weaned calves up at Parsonage. She looks stronger and her cough is less frequent than before—there seems little point in keeping her in isolation any longer.

It is colder again. There was only a gentle frost last night, but by mid-afternoon while David was out beagling it was freezing hard in spite of bright sunshine.

MONDAY 3RD

I feel sure now that Poogly's calving date was 3 weeks out. No matter how hard we look for promising signs of calving, none is apparent.

John seems happy to be back doing his yard work again. The knee stood up well to a good jolting when he was feeding up the cattle this morning.

LAMBING

Lambing, like harvest, is one of the high points of a farming year. The work load on a farm is not spread evenly. This is one of the periods of intense activity and hard work.

Lambing is close now. I was out with Graham on Burrow Brimclose this morning and there is no doubt some of them look heavy enough to lamb in a couple of weeks. We will be lambing later than usual this year and everyone is taking a deep breath before it all starts, because once it does start it seems to go on for ever. Ideally we would have our lambs born healthy, between 7 in the morning and 7 in the evening and in a cold but temperate spell of weather. It does not happen. There are always problems, awkward births, disease and deaths. When you turn out yet again at 3 in the morning on a freezing windy night, it is a little difficult to be even-tempered with the milking cows 4 hours later down in the dairy. Everyone is involved, and it is not unusual to have a limp lamb in the bottom oven of the Rayburn in an effort to bring it back to life. Already Hettie has a line of feeding bottles, teats and milk powder on the window sill above the sink, all waiting.

TUESDAY 4TH

SLURRY

All fields are spread with slurry once every 3 years. Too much too often will sour the grass. This should only be done when rain is on the way, so that it soaks in well.

The slurry pit will need emptying and spreading fairly soon now. It is a small one for this many cows, and we shall need to empty it more often than we thought.

Graham went into market this morning, stopping on the way at Mr. Newcombe's butcher's shop in Hatherleigh. He took in two fat pigs for Mr. Newcombe to kill. In the market itself Judy's Friesian bull calf made only £50, while the first Hereford cross heifer made only £30. The market has slumped badly after Christmas, and the fat cattle market prices were way down. Usually they rise after Christmas, and John had kept his four best steers back waiting for prices to rise, but they have gone right back. There is a lot of Irish beef come on to the market, and the demand from E.E.C. countries for our calves has suddenly vanished.

WEDNESDAY 5TH

Emily has settled in well with the calves in the shippen, but her ringworm has re-emerged.

A milder day, so David went out and cut 2 days' worth of kale. It could be frosty again tomorrow and then he would not be able to cut it.

Graham went to a Young Farmers' party at Belstone this

Steers in the shippen

evening and got back to discover Lavender had calved. She's the third of those four heifers and, like the others, she had been running with a Hereford bull. The calf is a heifer, a Hereford cross Friesian heifer, but with a difference. All Hereford cross Friesians have white faces and most are black elsewhere, but this cross occasionally produces a red and white calf—it's brown really but we call it red. And so this one is, almost fawn and white—very pretty but we would have preferred a black and white bull calf. Strangely enough farmers do not much like the red ones, so that means she'll fetch less money.

THURSDAY 6TH

A day of cleaning out shippens and dung spreading. Emily's house had to be scooped out. As with all cattle houses we only clean them out occasionally and there is a good reason for that. Cattle lie down for much of the time they are inside and it is important they lie on something warm. So instead of scooping out each day and putting a thin layer of straw down on cold concrete, we let the dung build up like a mattress, spreading a dry layer of straw on top each evening.

David took several loads of dung up to Berry, but it's a slow process. One spreader load seems to go nowhere; maybe two or three rows up there, that's all.

The new calf looks skippy, and adds some colour to the exclusively black and white calves down in the calf pens.

FRIDAY 7TH

The muck down at Burrow is appalling. We can't use the scoop and loader to get rid of it because the digging up of the road down there has loosened the stones, so we walk around in a pool of mud and manure. We hope to be in the new parlour by next week and the cows can then go back in the yard on the concrete. Then we will get down there with shovels and clear it away.

SATURDAY 8TH

The milk yield has dropped. Poppy and Angelina had been giving noticeably less milk recently and this morning Graham felt the tell-tale hardness in their udders. He tested further, stripping the milk into a strip cup, and his suspicions were confirmed. Both cows have mastitis in one quarter. He inserted the mastitis tubes to clear it up.

Down at Burrow there is only half a bay of hay left in the Dutch barn and after that only the three bays of the barn up at Berry—there are about 400 bales to a bay. That's all there is for the milking herd, and although there are still a few bales left in the big barn at Parsonage, it is going to be very tight. Every bale counts. Hay will be expensive in the early spring.

The milking parlour is crawling with builders and electricians and plumbers and fitters, all hammering away to get it all finished by next week. And it will be so comfortable to be milking inside again.

SUNDAY 9TH

There's snow all around the country and over most of Devon, but it has not reached us yet. Dartmoor looks more like the Alps, the white slopes bringing it closer somehow.

The rams have been moved into another field on their own

whilst the ewes are lambing. They have been breaking out. The old ram has a nasty swelling on his face and it burst this morning. Graham has painted the wound with antiseptic, so that the ram looks as if he's got a black eye.

Graham and I went fencing in Lawn Field this afternoon in driving rain. All the fields we will need for ewes and lambs are now fenced and ready.

RAMS
Often difficult to contain, they will dislike the initial separation from their ewes. When tupping is over, complete separation has to be ensured otherwise lambs could be born at the wrong time of year.

MONDAY 10TH
Graham took in twelve of the remaining hoggets to the fat market in Hatherleigh. They made nearly £25 each—about the best price of any we've sold. They have started weighing everything in kilos in the market and I can see it will take some time to become familiar with it. But we will go on talking in pounds, and hundredweights and scores until we get used to the other way.

More snow about and it has been threatening to fall over Iddesleigh all day. But so far it is only a threat. Eight miles away they had 6 inches during the night. It is coming though, and we must get that hay back from Berry before it does. We'll never make it up Berry Hill with snow on the road.

The old boar appears to be getting vicious in his dotage. John had put some of the heifers out in the orchard to graze and the old boar was out there with them. When John went to bring the heifers back in after tea, one of them had a gash inches long in her leg. We have never known the boar do that before—up to now he's been quiet and placid. It is the sows that can turn nasty when they have just farrowed. The heifer is limping a bit, but John had a good look at the wound and thinks she will heal up in a week or so. It was bad timing, as the heifer is due to calve in a couple of days.

David has brought in a lot of kale, just in case the snow comes.

TUESDAY 11TH
Bluebell, the heifer bitten by the boar, calved during last night—a Hereford cross Friesian heifer calf. She is a fair looking calf and Bluebell calved easily and unassisted. She seems well in spite of the gash on her leg, and there's a good udder on her for a heifer.

Snow came in the afternoon, and the sky is full of it. It came in from the moor in grey waves of mist cutting out the sun, and then moved away to the north.

Poogly is looking as if she might calve quite soon. She is heavy enough below, so it seems we were 3 weeks out in our calculations.

Parsonage Farm under snow

WEDNESDAY 12TH

By this evening we have had several snowfalls. It looks lovely, particularly when it is fresh with the orange glow of the early morning sun—but that is from inside looking out. Every bowl in the cattle yard needs thawing out with hot water. The lanes are like glass and the Mini froze so hard we couldn't even open the doors. But there are compensations. The world has turned white around us, and suddenly there is no more mud or mess.

The milking parlour is nearly ready—just the finishing touches. It has been like a jigsaw puzzle, and like all jigsaw puzzles there have been pieces missing from time to time. It is a web of pipes and wires and jars and funnels and hoses. Strung together in the right order they make a modern herringbone milking parlour.

The hens don't seem to mind the snow. There were sixteen eggs today, the best of the winter. The days are longer now already—it is not dark until after 5 o'clock.

THURSDAY 13TH

The fitter for the milking parlour was snowed in on Exmoor, so D-day for the new parlour is postponed till Saturday. It is depressing for Graham and David. Already their hands have taken on their winter redness and cracks, and neither can wait until they are back inside and washing down the udders with warm water.

We moved the sheep over to Brinnen in the afternoon, and they should stay there now till lambing starts. Our biggest field, it is an ideal sheep field with shelter and water down by the brook, and a long dry hillside for grazing. Graham feeds them mid-morning, taking over the hay and the sheep nuts in the link box. Every morning he checks them to see how near they are to lambing.

FRIDAY 14TH

The worst day of the winter, horizontal rain driven by gale force winds that brought in snow and sleet as well. David managed to get into Hatherleigh to bring back more sand for the cow cubicles, but it was too wet to unload it. Graham forced himself out to get the kale cut, but it was too muddy to get much done.

The dairy cows hate this weather. Even when they are lying in the cubicles the rain seems to drive in after them. But the fitter came to finish off the parlour, so at least we shall be bringing them inside tomorrow to milk in the dry. The first night in the new parlour is bound to be tricky. Cows are creatures of habit and routine, and I can think of a few that might not prove too co-operative.

Two of our ducks were taken during last night; the fox no doubt. I found a trail of white feathers all the way up from the pond across Lawn Field and up into the woods beyond.

SATURDAY 15TH

More driving snow and sleet, just the day not to begin working in the new parlour—but this evening it was ready and finished and David and Graham used it for the first time. They started milking around 6 o'clock and got back home at 11.45 p.m.—a little tired. It was not just that the cows were difficult to get in, but Graham and David had to get to know the new machinery as well. And, of course, the machine itself had teething problems—some of the feeders were not working, and one of the clusters seemed to be out of rhythm.

Although the bulk tank is installed we can't use it until there is hard concrete outside the dairy for the tanker. It is mud and snow at the moment. We hope to start the concrete next week. Meanwhile we are still letting the milk into churns which have to be driven up to the cross-roads to the milk stand as usual.

SUNDAY 16TH

A clear cold morning and Dartmoor has come out of the clouds. Milking is already taking half the time when we started yesterday. We began at 7.30 this morning and we were back home for breakfast by 10.30. The cows came in better, and the knobs and buttons seemed to be coming to hand more easily. Another day or so and it should be quicker than with the old bail.

The sun shone warm this afternoon and I went down to have a look at the Torridge below the Marsh. It was bank high and roaring brown.

MONDAY 17TH

John drove into Holsworthy this morning to order a new yard scraper for the cow yard and a new link box. The yard scraper has been repaired several times by Mr. Vanstone, the blacksmith at Monkokehampton, but there is just not enough sound metal to join together any more. It was eaten through with rust and there has been clear daylight through the metal now for many years. The link box is in constant daily use all the year round, but the old one was too small and too battered to cope with the job. Hitched up to the tractor the new one will be able to take a hundredweight of sheep nuts and twenty bales of hay.

Milking becomes a smoother operation with each session, although a few of the younger cows are still nervous. Herma came in easily enough this morning, but decided she didn't like what she saw, so she walked right along, pushed up the iron gates at the far end and walked out.

One of the sows took a bite at me this evening during feeding. They get very excited and noisy as they are being fed, and she reached out and tore my coat. Food is the only way to quieten them down.

LINK BOX

Fixed hydraulically to the tractor, it can carry hay bales, small animals or tools around the farm.

SCRAPER

Linked hydraulically to the rear of the tractor, the scraper is a broad sheet of metal with a rubber flap at the bottom. Scraped along the ground it saves much of the spadework in cleaning down yards and cattle sheds. Used daily in the cow yard.

TUESDAY 18TH

There was a new calf in the shippen this morning. Violet, who was not due for another week, had calved down a lovely, long-legged Friesian calf and a heifer at that. Violet herself is a good milker, just the kind of cow we need heifers from to build up the herd.

Hatherleigh store market. David took in two of the Hereford cross Friesian calves, the little red one and one of the black ones. But prices were disappointing yet again. The red one was small and buyers don't like the colour, she made only £17. The other made £35, but that is not much compared to last year's prices.

The milk yield is up to over 114 gallons now—the freshly calved heifers are helping and every milking sees the cows

MARKET

Hatherleigh, the nearest market, is 3 miles away. Each week there are two markets. Monday is the fat market, when all the fat animals are sold—that is animals fit and ready to kill. Tuesday is the store market, when animals are sold to be kept on. The market takes a 2½ per cent commission of everything sold, and sale is by auction. Farmers go primarily to buy and sell, but it is also a day for commercial discussion, to consider price trends and market activity.

more relaxed. There is still mud everywhere around the yard at Burrow, and we can't have any concrete laid whilst there is still a chance of frost.

We have bought a new cockerel to replace the one that died last year. He's a Maran, not yet a year old, and a huge bird. We expected him to stamp his authority over his hens at once, but at present all thirty-three of them bully him mercilessly, chase him off food and peck at him whenever they can get close enough. He's a fine cockerel with a blood-red comb over a ruff of light grey feathers.

WEDNESDAY 19TH

Poogly has done it at last. First thing this morning, John noticed her udder was swollen to huge pink proportions, and he thought then she would calve sometime during the day. Graham went over after breakfast and took her into the big barn with Violet—there's more room there. By this time she was showing a bag, and when she had licked herself over she rustled around in the fresh straw looking for the best place to lie down. She found it and settled. She was about to calve when Violet's calf came over to have a look, followed closely by Violet herself. This was too much for Poogly who got back on to her feet and began to chew the cud again. John did not like the delay, so we drove her back into the shippen again, she calved 10 minutes later. It is a fine Devon cross Jersey bull calf—Bournville brown with a hint of blue in his nose. Poogly took to him at once and licked him from end to end several times over. Three-quarters of an hour later she was still at it, knocking him over with her affection as he staggered about the shippen.

Graham brought the sheep back from Brinnen to pick out the early lambers. He has kept the sixty ewes that look close to lambing out on Furze Close and driven the others back to Brinnen. They are due to lamb on the 21st, so it could be any day now.

THURSDAY 20TH

The man from the Ministry came to ear punch the calves for the subsidy. He clips the ear just as they used to clip tickets on the London buses. The calves had to be forced into the crush, but once inside made little resistance.

David milked by himself in the new parlour, his first solo. It went well and he finished early. Even Poogly came in easily. David brings her into the calf pen to feed her calf morning and evening, but meantime she is having to readjust socially in the herd. Emma seems to have taken up with Hyacinth and Poogly was standing alone in the yard this evening. She is content to be back with her calf, although the calf doesn't seem to know what to do and David has to remind him.

But the best news of the day must go to the sheep. By this evening, two Suffolks had lambed—a double each. So they've started. From now on it will be every day and night for over a month. It will mean late nights, and early mornings. The lambs will have a life of between 4 months and a year before slaughter—except, of course, for the select few ewe lambs we keep back for breeding.

The Government have put £3.50 subsidy on the price of each pig in an attempt to get the pig trade out of the doldrums. It will help of course, but with the cost of feed accelerating all the time it won't help for long.

SUBSIDY

For each calf the farmer has reared to 8 months, except dairy heifer calves, the Ministry of Agriculture gives him a subsidy of £9. The ear is punched so that the subsidy is not given out twice on the same calf. Once punched the ear does not grow back again.

FRIDAY 21ST

Our first lambing disappointment. David went out to check the flock on Furze Close before milking early this morning. He found a Suffolk ewe licking over her three dead lambs. Two of the three looked big enough to have lived, and it was not that cold a night. It does happen with triplets that there is a weakness in all three. Perhaps we could have done something to help if we had discovered them earlier, but born as they were in the middle of the night there was really nothing we could have done—short of sitting out all night with the flock. It's a blow coming so early in lambing. The ewe is well enough though and John bought in an orphan lamb to suckle her. He penned them in together, tied up the ewe and left them. By this evening the ewe was happy to let the lamb suckle.

I saw a heron rising off the Lawn Field pond this afternoon—a slow flight of great elegance and power.

LAMB DOUBLES

Not uncommon in a Suffolk flock, the doubles have a high rate of survival. Each ewe is carefully checked to ensure there is milk from both teats. If not one lamb has to be taken away, and either brought up by a ewe with only one lamb or fed on the bottle.

Triplets are much less common, and with a poorer survival rate. They are generally weaker lambs.

SATURDAY 22ND

Just the one ewe lambed this afternoon, a single. Both look well. Once they have lambed Graham brings them away from the others and on to Front Meadow which is nearer the house and where there is some fresh grass.

Graham took the tractor over to Holsworthy to fetch the new scraper and the link box. The link box is huge, twice the size of the old one.

The new scraper went into use right away, and Graham spent all the afternoon cleaning up the yard and the lane at Burrow. Much of it had to be spade work because the lane is so rough, but now we are all ready for concreting next week—if the frost keeps off.

SUNDAY 23RD

A rise in the milk yield. There are good reasons for it: the newly calved heifers and cows are coming into the full flush of milk, and we have fewer cows drying off.

One double and a single born during the day, and all of them are strong, healthy lambs. John, Graham and David share the lambing between them. David checks them first thing in the morning before milking, and he looks for any lambs born during the night or any ewes about to lamb. He moves the new-born lambs over to Front Meadow and goes off down to milking. Graham and John go out again after breakfast and again around lunchtime, looking always for the ewes sheltering up against the hedges, on their own away from the flock. Again John will check them around teatime, and then Graham will go out last thing at night, driving around the fields in the tractor, using the headlights to search out any new-born lambs or ewes that could be lambing shortly. It is not as regular as this implies—but there is always someone out checking the flock every 3 hours or so throughout the day and into the night.

LAMBING
Even when lambing inside with ewes all around, a ewe will always go off by herself to lamb away from the others. Like cows they seem to like to be alone when giving birth.

MONDAY 24TH

David was up late last night waiting for a ewe to lamb. She had already had one but it was clear there was another on the way, so David waited with her in the Big Barn to make sure she could manage by herself. She did, but she took her time, and David clambered into bed after midnight.

There was another double born to a Suffolk ewe this afternoon, so we are up to a dozen now, and most of those are doubles. Of course heavy lambing can produce problems—as we had the other day with those triplets—but by and large the lambs are sturdy enough to do well. All the ewes that have lambed so far have been Suffolks.

SUFFOLK SHEEP
A much favoured breed. They lamb well. The Suffolk ewe produces an average 1.5 lambs in a season. The fleece of a Suffolk grades well and the lambs grow fast with deep bodies. They have a lean carcass.

The evenings are lengthening noticeably. It is not really dark now until well after 5.30 and with a lot of work to get through, during the day that extra hour or so of light makes life a lot easier.

TUESDAY 25TH

A terrible night with the most violent winds of the winter and driving rain. We were up all night at intervals checking on the sheep. John was worried that any lamb born out in gales like that might not find shelter enough to survive, but all was well by the morning and the storm had passed on leaving high winds and racing clouds.

There were six singles and a double born during the day and this evening Graham has brought all the ewes with lambs back into the Big Barn for shelter.

The wind has got up again and now it's blowing a gale. There are branches down everywhere, and we expect power cuts any time. The candles are ready, the oil lamps filled and trimmed.

WEDNESDAY 26TH

Three doubles and a single by this evening, but one lamb looks unlikely to survive. He's small with a distended stomach—it looks as if he may have a blockage of some kind. Graham has tried a dose of liquid paraffin, but it does not look promising. Sheep are strange animals—once they are ill, they cab away. You can try everything and anything, but once a lamb begins to deteriorate there is very little one can do.

CAB AWAY
Sheep seem to fade away when they are sick. They have little resistance to pain or illness.

We cleared out one side of the Big Barn this morning to make room for the lambing ewes. Graham has divided the side into two parts—one for those that have lambed already, and one for those about to lamb. With the nights as wild as they have been we shall be bringing them in each evening now, and putting them out by day if the weather is fair. We do not like to house sheep for too long. The more they stay out the healthier they will be.

In the afternoon Graham spread a load of dung on Burrow Brimclose. He will be concentrating now on fertilising the Burrow and Redlands fields. These are the main summer grazing fields for the herd and we will be taking a crop of hay off one Redlands field, so the more we feed the grass at this stage the better.

David and I have been managing the milking for a couple of days, while Graham sees to the sheep up at Parsonage. Graham says this may be the reason the milk yield has dropped.

THURSDAY 27TH

A lamb born on Wednesday died during the night, and I think there's another one that will go the same way. During the day there were two Suffolk singles and a double from a Dorset ewe. David helped the Dorset to lamb, pulling off the second one himself. The ewe seemed unaware there was another one there—she was so busy licking over the first.

CONCRETE
The ideal surface for farm yards and lanes. It can be cleaned off easily with the scraper, and when laid properly (6 inches thick) will withstand the weight of tractors and milk tankers. When being laid it should be ridged so that it is not slippery for the stock.

One of the singles looks weak, he can't seem to lift up his head and finds it impossible to suck his mother. Of course, we can hold him up to her, but in the end he has to manage by himself. He's not likely to survive.

The first concrete is down outside the parlour. It will not be long now until we go over to the bulk tank.

BLOWING CAKE
The cake lofts are situated in the roof of the new parlour. They contain 8 tons of cake and this has to be forced up from the bulk feed lorry through a long feed pipe which is attached to an inlet hole at the end of the lofts. Delivery is once every 5 weeks usually.

The rams were moved from the orchard on to Lower Redlands. They kept breaking out to join the ewes so they've been banished to the other side of the road.

We are having some difficulty in blowing the cow cake up

into the lofts above the new parlour. It will have to be fixed before the next delivery.

I was taking the horses their hay just before dark when I saw the first snowdrops along the banks on Burrow Meadow.

FRIDAY 28TH

As the lambing increases Graham is putting in more time with the sheep, leaving most of the dairy work to David and myself. There were ten lambs born today, the last single keeping Graham up till well into the small hours. No more casualties, but it is becoming clear that many of the doubles are not as strong as they might be. John feels sure it must have something to do with the long dry summer when the grass was poor and lacking in nutrition for the ewes. Then since the autumn, there has been this continuous heavy rain and sheep don't like that either. All the singles look strong enough though. Extra corn for the in-lamb ewes will be necessary from now on.

Marigold has some mastitis in one quarter. It is not bad, but David had to treat it. David seems to have a natural affinity with the cows. He only has to stroke a cow and talk to her quietly and she calms down, and when the cow has kicked the cluster off for the fifth time, that needs patience. His favourite is Herma—the one that kicked his tooth out last year.

QUARTER

The udder has four quarters, one teat to each. Mastitis usually occurs in only one quarter.

SATURDAY 29TH

Fourteen more lambs by nightfall, and John is out with Graham keeping watch in the Big Barn. It's a cold night, the rain giving way to clear skies and a sharp frost. We brought the ewes back in early this afternoon in case it froze, and it has.

Three of yesterday's lambs are looking poorly. John doesn't think they have much chance but we have to try. We've brought them in under the warmth of the lamp and Graham is feeding them on the bottle; and it's five times a day when they are this young. There is nothing specific that we can treat—they are just weak lambs. One of them has no interest even in the bottle.

SUNDAY 30TH

Still cold and bright under a pale sun—good weather for lambing. We put all the ewes and lambs out, except for the three still struggling for life under the lamp. The two bigger ones look stronger this morning, but the last one seems even more feeble.

Only three more born today, a double and a single. In the afternoon Graham brought back more ewes into Furze Close

Opposite:
Feeding an orphan lamb

Graham delivers a lamb

Delivered safely

from Brinnen, so that there are nearly eighty ewes on Furze Close now that will be lambing soon, too many to bring inside. He has begun to spread the couples—ewes with their lambs that is—out on to Lawn Field and Upper Redlands, leaving Front Meadow clear for the new-born lambs and ewes. The bleating everywhere is continuous, except at night. The lambs stay close to their mothers at first, but as confidence grows and as the ewes become less protective, they are for ever getting muddled up and losing each other and that's when the noise starts. At feeding time in particular when the ewes busy themselves over the troughs, the lambs set up a chorus of bleating until the mothers return to them, sniffing each one in turn till they find their own.

MONDAY 31ST

Seventeen lambs born today, most of these doubles. But there have been more problems. One of a triplet born in the early hours was dead at birth, and the other two look weak.

This afternoon Graham was out on Front Meadow with a ewe that was about to lamb. He was concerned because she had been looking unwell for a day or so. He pulled the two lambs off her, but she seemed unwilling or unable to stand up to clean them off. She died shortly after the birth—Graham thinks it was some kind of inflammation. She was a two-tooth and the loss is bitter. Her lambs are strong though and Graham brought them in under the lamp and will be feeding them from the bottle—our first orphan lambs. It will be a day or two until they get used to the bottle—half a bottle five times a day. We mix up powdered milk with warm water and then hold the teat inside the lamb's mouth until he understands what to do.

TWO-TOOTH

The age of sheep can be gauged by counting the number of teeth. A Two-tooth would be 1½–2½ years old. A Four-tooth would be 2½–3½ years old, and a Six-tooth, 3½–4½ years old. After 4½ years they have a full mouth. Sheep have front teeth only at the bottom. Born with eight baby teeth, two new broad teeth take the place of the baby teeth each year.

David sent away over 115 gallons. We are becoming really swift in the parlour.

John was up until 3 a.m., so officially it should come in tomorrow's episode, but I'm writing this at 3.15 a.m.—it feels like the end of today, not the beginning of tomorrow.

There have been problems, but lambing has begun satisfactorily with the early lambs growing on well.

The following cows have been inseminated during this month:

Lucy—Thursday 20th
Herma—Thursday 27th
Primrose—Thursday 27th

COUPLES UNDER COVER

The ewes are in the shed
Under clapping wings of corrugated iron
Where entering rays of snow cut horizontal
Fiery and radio-active, a star-dust.
The oaks outside, half-digested
With a writhing white fire-snow off the hill-field
Burning to frails of charcoal
Roar blind, and swing blindly, a hill-top
Helpless self-defence. Snow
Is easing them, whitening blanks
Against a dirty whiteness. The new jolly lambs
Are pleased with their nursery. A few cavorts
Keep trying their hind-legs—up and a twist,
So they stagger back to balance, bewildered
By the life that's working at them. Heads, safer,
Home in on udders, under-groin, hot flesh-tent,
Hide eyes in muggy snugness. The ewes can't settle,
Heads bony and ratty with anxiety,
Keyed to every wind-shift, light-footed
To leap clear when the hill-top
Starts to peel off, or those tortured tree-oceans
Come blundering through the old stonework.
They don't appreciate the comfort.
They'd as soon be in midfield suffering
The twenty mile snow-gale of unprotection,
Ice-balls anaesthetising their back-end blood-tatters,
Watching and worrying while a lamb grows stranger—
A rumpy humped skinned-looking rabbit
Whose hunger no longer works. One day
Of slightly unnatural natural comfort, and the lambs
Will toss out into the snow, imperishable
Like trawlers, bobbing in gangs, while the world
Welters unconscious into whiteness.

FEBRUARY

TUESDAY 1ST

Another heavy lambing day with fifteen born by 10.30 tonight. John was talking to other farmers in the market this morning and there appears to be a weakness in many of the doubles. It means we can take no risks, that if possible we should be there at every birth to bring in the weak lambs and keep them in the warm and dry until they are strong enough to fend for themselves.

One of a double born late this evening is looking poorly and it is doubtful if he will survive till morning.

Graham has to inject every lamb against lamb dysentery. It must be done within 12 hours of birth and has to be done if the lamb is to have a chance of survival.

Mr. Byrne, the vet, came this afternoon to give the cows their T.T. test. David drove the herd through the milking parlour as if for milking and the vet checked each one for skin thickness. He injected them with tuberculin, and any that have a positive rection when he comes back on Friday will have to be slaughtered. But that is most unlikely.

LAMB DYSENTERY

Enteritis of the small intestine, leading to scour. Every lamb is injected at birth to prevent infection.

T.T. TESTING

Every 2 years there is a mandatory T. T. (Tuberculin Test) on every heifer and cow in the country. A vet is sent by the Ministry of Agriculture to test the cows to prevent the spread of T.B.

WEDNESDAY 2ND

Not so many today, just three doubles and a single, but two of the doubles lambed well into the night. It was David's turn to stay up to look after them. He was not in bed till 2.30.

The one lamb that was sickening with the scour yesterday died in the night, and there's another today that is looking unwell. Graham spread some paste on to its tongue and providing it is not too far advanced and the lamb is still strong enough, there should be a good chance of recovery.

When first born the new-born lambs and ewes are put out close to the house on Front Meadow, then they are split into singles and doubles and taken away to Upper and Lower Redlands. But with the lambs coming as fast as they have been we needed now to make more room close to the house. So our earliest doubles and singles have been taken up to Berry in the cattle transporter, leaving Front Meadow clear for those about to lamb. There is good shelter up there and most of those moved are a week old at least so they should be all right. They will need checking regularly, once or twice every day.

Pruning hedge trees

BULK TANK

A refrigerated tank that cools the milk down to 40°F and stores it until collected by the tanker. It has a capacity of 240 gallons. The driver first measures and records the amount of milk in the tank. Then the milk is sucked through a pipe from the bulk tank into the bulk tanker. Before leaving he sets the automatic washer which cleans the bulk tank thoroughly before the next milking.

The bulk tank was connected up today, so we will be having the lorry come down to us from next Monday. The concrete is all dry so there should be no problems.

A fierce wind blew up around dusk bringing down branches everywhere. Graham had to get out at 4 in the morning to bring in any new-born lambs.

THURSDAY 3RD

There are fifteen lambs by this evening, most of them doubles. One lamb died shortly after birth, just too weak to live.

A lot of scraping and shovelling today. Both the sides in the Big Barn needed a good scrape as did the lower yard outside. The lambing has left us little time for this. The pile of dung is already half-way up the Old Barn wall; we shall have to spread when we can but it is too wet at the moment.

Down in the dairy, Hyacinth has mastitis in the same quarter as before. It is an ominous sign. She is a sixth calver, and this is the second time she has had mastitis in this lactation. It is something older cows are prone to. She may be coming to the end of her useful life. The decision will be a difficult one—she has been with us for some time.

FRIDAY 4TH

Only three doubles and a single, but two of the doubles came after 10.30 at night and David had to wait up till the early hours again.

John had to get Graham up in the middle of last night to help him pull off a lamb that seemed stuck. The head was out but the ewe was very tight. The lamb is huge with an oversized head and looked unlikely to live. But this evening he is looking stronger.

A tragedy this afternoon. There was a ewe John was sure should have lambed—indeed the afterbirth was hanging out, but there were no lambs. It was clear they were either lost or inside still. John caught her and took off two dead lambs, but the ewe is all right.

The vet came again to check the results of his tests on the herd. All's well and we are officially T.T. tested now. But he so upset the herd by putting them all through the milking parlour for their examination that they behaved abominably at milking this evening, refusing to come in, kicking off clusters and messing everywhere.

Hyacinth is a bit better after treatment. And Poogly is at last beginning to stamp her authority on someone—a Friesian heifer newly calved. She has found someone she can bully at long last.

Everyone is very tired. And we are a long way from the end of lambing.

SATURDAY 5TH

A day of cattle problems. It was a fine morning and John went back into the barn to drive the steers and heifers out into the fields. Somehow one of them had managed to lodge herself in the trough under the hay racks and was stuck fast. It took four of us pulling for fully 10 minutes to prise her out.

STEERS
Castrated males reared for beef.

It was Shamrock's heifer calf, one of our best. It was pure chance John went in there—given a rainy day they would have stayed in and she would have been stuck there all day, with very little chance of survival.

Then the cows were seen out in Mr. Rafferty's garden down at Burrow. David and I went down to discover all thirty-eight of them in his tiny garden. Lawns, flower beds, vegetable patch were all flattened. We drove them back into the yard easily enough. Somehow, the door from the yard had become unbolted and the wind had done the rest. It is very strange because no one has used the door for years. The only possible theory—hobgoblins apart—is that the cows' continual bumping against it must have loosened the bolt gradually, because nothing was broken. We shall have to claim the insurance for Mr. Rafferty.

Lambing is still at a peak, with ten today, four doubles and two singles, and no more casualties. The lambs seem to be stronger now.

SUNDAY 6TH

We have lost count some time ago now, but Graham thinks there are over 100 lambs already. There were three doubles and a single by this evening, but one of the doubles was a disappointment. Graham pulled off the first lamb and this was barely living, the second was dead at birth. We are trying to save the first one, but it will be an uphill struggle. He has taken to the bottle, but looks weak this evening. Every lamb is important and once we start trying to save one it becomes a matter of pride to succeed.

Bounce is being taught some manners. Till now he has been impetuous and tactless with the sheep, but a ewe with lambs at foot will turn on a dog, and he is now learning the subtler technique of moving the ewes without the aggression that would make them stop and turn.

The churns have gone. This evening we milked into the new bulk tank, with the milk being pumped directly from the collecting jar in the parlour straight out into the refrigerated bulk tank in the dairy. The 50 or so gallons we milked this evening look very unimpressive in the bottom of the 240 gallon tank. But it means no more carting churns about, no more changing churns during milking, no more labelling each individual churn. Tomorrow the lorry will come, record the level on the dipstick, pump it all out, turn on the automatic washer and that will be that.

MONDAY 7TH

Fat market and the last of the older lambs from last March—now "hoggets"—were taken in, along with the old ram. Four fat pigs went with them; there is that £3.50 subsidy on each, so it won't be too bad. We tried to drive the old boar up with them, but he wouldn't go. Next time we will win. Graham took them all in and left them. There's no time now to wait around on market days. Everyone looks the same, lamb-shocked and grey.

The bulk tanker came down Burrow Lane for the first time and pumped out our 115 gallons.

Another shift round of the ewes and lambs this afternoon, with twenty more singles going off to Berry to join the others already there, and these have come on very well. A slight lull in the lambing last night—and everyone had a good night's sleep for the first time since it all began. But lambing lulls never last.

The banks and hedges are full of snowdrops now. There are crocuses under the trees and small swelling buds on the branches. The fresh green shoots of daffodils are already coming through and it's mild and damp everywhere, the dead of winter seems behind us now.

Snowdrops by the Torridge

TUESDAY 8TH

Store market. Poogly's calf was taken in along with one of the Hereford cross Friesian calves. Poogly's made £25 and the Hereford cross over £50, so the market seems to have picked up.

David went in with them and found out what the hoggets and the ram made yesterday. The ram made £30 almost and the best hoggets the same.

Graham and David changed roles today. Graham had had his fill of sheep and took a 'holiday' down in the dairy.

Ten more lambs today, and another triple amongst them. David has brought them in on the front lawn with the ewe. They are well enough, but one is that much weaker. He's helping it along with the bottle.

Hettie had a chase around Front Meadow this afternoon. She spotted a ewe with a lamb's head hanging out, but the ewe was paying more attention to two other lambs that were not hers. Hettie called David out and they managed to catch her and pull the lamb off safely.

WEDNESDAY 9TH

The hay stocks are looking very low. With at least 2 months before any grass comes through there are only the three bays in the barn at Berry and one side of the Big Barn left. John wants to buy a few more young cows to put into the herd and he would like to keep on a few good beef calves, but there just is not the hay to go round. We may have to buy some more anyway, but with hay at around £55 a ton, we certainly do not want to.

BEEF CALVES
Usually bull calves if they are Friesian, but heifer crosses are also kept for beef. Among the beef breeds in this area are: North Devon, South Devon and Hereford.

Emma looks likely to calve at any time. She is very wide, and her udder is fully sprung already. We are checking her every 2 or 3 hours.

Four doubles and a single by this evening, and all are well. The triple on the front lawn is doing well now, with the smallest one a lot stronger. But it will be a rough night for the lambs. It started raining soon after dark and it's lashing down outside now. No young animal likes it so wet and we shall have to be out there early tomorrow morning to bring in any saturated lambs.

THURSDAY 10TH

The rain drove down all night long, and this morning Graham went out and rescued two washed-out lambs on Front Meadow. He brought them in and put them under the lamp in the pig sty. They should be all right.

There were three singles and a double born today, but the ewe with the double is giving milk only on one side so we have had to take one lamb away for a few hours. When you take a lamb away from its mother for more than 6 hours or so it is doubtful if she will accept it again, especially if she has another lamb already.

A buyer came for the old milking bail today and towed it away on a trailer up Burrow Lane. Neither David nor Graham were sorry to see it go. It held memories of cold, windy nights in December and frosty mornings in January.

EMERGENCY BULK TANK

Burrow Lane is steep and in snow the milk tanker would find it impossible to make it to the top. The emergency tank can be taken up to the main road on the link box to await the tanker. This procedure could be continued indefinitely if necessary. The milk would be pumped each day from the bulk tank into the emergency bulk tank, leaving the bulk tank empty for the next day's milking. If the tractor could not get up the lane then after 3 days or so the milk would have to be poured away. This eventuality is very unlikely.

FRIDAY 11TH

Graham thinks there have been 155 lambs born and still living by this evening, and we estimate we shall finish with around 250 in all.

Our emergency bulk tank arrived this morning. It will take 270 gallons and can be fitted on to the link box. If it ever happens that the lorry cannot make it down Burrow Lane, or the lorry fails to come one day, we now have storage for 510 gallons of milk—sufficient we hope for any emergency.

Catkins are out everywhere along the river. The land is beginning to breathe again.

SATURDAY 12TH

We shall keep fewer sheep next year. With the increased number of milking cows we feel we have too many sheep for our acreage of grass.

SHEEP GRAZING

"A sheep's worst enemy is another sheep." Overcrowding the pastures with sheep contaminates the ground. Farmers describe such ground as "sheep sick". Sheep should be moved on often to clean pastures.

One of the triplets born earlier in the week died in the night. We all thought he would live. Each lamb that dies is a sharp disappointment. Everyone is tired during lambing, but in spite of this there is a sense of continuing anticipation and involvement with every ewe that lambs.

Our Jerseys are riding high. Poogly is giving over 4 gallons of milk a day—2½ in the morning, 1½ in the evening. And Emma is looking close to calving.

SUNDAY 13TH

Emma has surpassed herself. A lovely light fawn heifer calf was born this morning at 7 o'clock. Mother and daughter are fit and well. She is a tall, strong calf and will be a great asset to the herd in 2½ years' time.

Ironically, Emma gave birth on the day John noticed that Emily is poorly. Emily, whom we thought recovered had seemed increasingly jaded alongside the other calves. She was off her feed and coughing badly again. We brought her back into isolation and called the vet. She was discovered to have a temperature of 105 and to be breathing only with great difficulty. He diagnosed pneumonia at once and injected her. If she responds overnight then we have a chance of saving her. We covered her in a blanket of straw and left her. Tomorrow will tell.

PNEUMONIA

Inflammation of the lungs. Virus pneumonia is brought on by overcrowding and poor ventilation. Weaker calves are particularly susceptible to any form of pneumonia. An antibiotic injection relieves the condition within 24 hours.

Graham has put some sheep down on the Marsh. There is good grass down there and providing the river does not flood they should be there for some weeks now. He has split the flock into six separate groups now—two of them are in-lamb ewes, then there are the big singles, big doubles, small singles, small doubles. The bigger and stronger they are the farther away from the house they can be left.

MONDAY 14TH

Graham and David managed to drive the old boar into the trailer for market. He went up easily enough this time following one of the sows. She seems barren. She has been running with the boar for some time now with no results, so they went in together. The old boar made £40 and the sow £50. John and Graham are keeping their eyes open now for a good young boar.

Emily looks a great deal better this morning. The vet came to inject her again, and things look more hopeful. Her temperature is down now to 101, but the wheezing is still there. Her appetite is back and she is eating her hay and concentrates and drinking her water. This evening last thing she was chewing her cud—so there is some hope.

There were three doubles and a single today, and they are all well.

TUESDAY 15TH

A day of changing weather, beginning bright as a May morning but later raining so violently that Graham had to go down to the Marsh to check the river level. It was rising fast, but there are still a few feet to go before it will begin to flood over. The sheep are in no danger. There were four doubles and a single by this evening.

Emily is holding her own we think. The vet costs money every time he comes and unless she recovers well there really is no point in going on. We are giving her till Monday and between us we will decide then.

WEDNESDAY 16TH

Three doubles and a single born today. All the lambs look fit and well and there is no sign of the scour.

Heavy rain and wind most of the day, but there was fencing on Burrow Brimclose that had to be done; so David and Graham spent the afternoon hammering in wet stakes that the cattle had loosened by rubbing.

David has been having problems with the bulk tank; it seems to cool only spasmodically, so an electrician was called in to fix it.

THURSDAY 17TH

Just three singles this morning, and it seems there is another lull in the lambing, which is just as well with this weather.

John is battling through his farming accounts for the last financial year, surrounded by piles of bills, invoices and receipts. The great debate is how much to re-invest in machinery and buildings, how much to keep back for themselves, and how to avoid being hammered by the taxman. It is his least favourite side of farming.

Emily's looking up again—eating and drinking and interested and alert. Quite different.

FRIDAY 18TH

Graham went off early down to Marsh to check the river—it was bank high already and rising. So he brought away the ewes and lambs and put them out on Ferny Piece for safety's sake.

We had a grand clean-up down at Burrow. We can see stones now and feel them under our boots—quite reassuring after all that mud and muck.

Emma's new calf is taking to the bucket already. A calf's natural bent is to look up for her milk, not to seek from below. David has to force her head down into the bucket, and splash her mouth gently. She is getting the idea.

SATURDAY 19TH

Sun at last.

We collected 2 days' cutting of kale off Long Close. There is more than an acre left so there should be enough to last us till the end of March. David ran along a strip with the tractor and mower and we forked it up into the link box.

We had a mammoth clean-up in the lower yard, scooping up lakes of liquid mud and dung. It has been in corners for some time with a film of green slime lying on the surface.

No more lambs today. There are only fifty more ewes to lamb and some are not due until mid-March.

SUNDAY 20TH

A wet one again, but it dried up enough for us to pick up the kale in the afternoon. Unless we get some dry weather we shan't be able to get on to the wetter parts of the field to cut it.

A bull calf was born in the night to Rita. He's over 10 days early by our calculations, but he's a big enough calf and looks strong at this stage. He'll be sold in a week or so.

MONDAY 21ST

One double born 2 weeks premature died within hours of birth. They were very small and thin and there's little doubt the ewe was upset by the hail, rain and wind. We have put one of our bottle-fed orphan lambs in with the ewe and she has taken to her easily. Graham rubbed one of her dead lambs on the orphan to encourage the ewe to believe the lamb to be hers. The ewe was taken in at once. Very often we have to tie the ewe up for days in a pen to make her take a foster lamb.

The hunt met this morning at the Duke of York, Iddesleigh, a few horses and the hounds. They failed to find even a scent on our land but were more successful elsewhere. Two foxes were killed—but a vixen is still crying out in the dark this evening, so they missed one.

TUESDAY 22ND

It is warmer today in spite of the rain. The rooks are back in the trees by the lane, and the herons are nesting in the woods above the Torridge. There are daffodils out already on the banks and green shoots of life again on the twigs along the hedgerows. It is still light now at 6 o'clock and the hens are laying up to twenty eggs a day.

We have over 200 lambs now, with another five this afternoon.

BANKS

Constructed as field and road boundaries, they abound in wild flowers and shrubs. They provide shelter for the animals against the wind.

Winter trees

WEDNESDAY 23RD
A dry day, and David has begun emptying the slurry pit. It is full to overflowing now—it should have been done weeks ago, but it has been too wet. David thinks there may be as much as fifty loads to be taken out—he managed five of them by milking this evening and spread it out on Burrow Brimclose.

Graham has been taking advantage of the lambing lull to work on the fences around the farm. The earth is soft, so hammering in stakes is easy enough. Every year there seems as much to repair and replace, no matter how much we do.

No lambs today at all. We moved twenty-five of the ewes with singles on to Lawn Field this afternoon where there is some grass to pick. Right the way through now until April when the grass will be growing again, it is a struggle to find fresh grass, so we must keep moving them around every week.

THURSDAY 24TH
A lamb again. One was born during the night out on Front Meadow—a white-faced Border Leicester.

Emily improves with every day and is now back to her habit of mounting you when you go in to feed her.

Seven loads of slurry were spread out on Burrow Brimclose—all David's work. He says he can tell now that he has been working on it for 2 days, the level in the slurry pit has fallen by a few inches.

David cut kale enough for 2 days and we picked it up after him this afternoon in sunshine that felt warm on the back for the first time. The warmth is beginning to be felt elsewhere. I saw a bumble bee out in the kale field, and in the lane on the way back a pair of Brimstone butterflies, the first of the year.

BUTTERFLIES
In spring and summer butterflies abound along the banks and hedgerows. It is a good environment for butterflies in this part of Devon. There are many patches of waste land as well as the hedges.

FRIDAY 25TH
Rain again. Milking is a wet enough business as it is, but with the cows dripping with rain it is a lot worse. Still it cleans them off. We have laid on warm water now for udder washing and the cracks on David's hands have healed over already. The cows seem to appreciate it too—there is no shock spasm or kicking when we start the washing.

Ten loads of slurry were spread this afternoon, and David is planning on emptying it within 10 days or so.

We need a long dry spell now. We shall want to plough as soon as we get it so that we can till the corn in March. Any later and the chances of a good harvest are diminished.

SATURDAY 26TH
I wonder if we shall go on with pigs. We have worked it out

and there is no doubt we are making a loss on every pig we sell. It does not encourage John to go out and buy another boar.

The cows are all skippy down in the dairy and seem unsettled in the yard now. They can smell the grass. They are always looking longingly over the fence into their fields, but they will have to wait a while yet.

SUNDAY 27TH

Celandine's Guernsey calf will be moving in with Emily soon. She is not doing well in with the others up at Parsonage. The building is not ideal for calves—there's too much humidity under a low ceiling. Emily is in an unused horse's stable with good ventilation, and it is clear this has helped her on enormously.

There have been two whole days without rain and the ground is even dry in places. We will be ploughing soon now.

MONDAY 28TH

Phyllis, our slowest milker, has caught a chill and David drenched her after milking this morning. He noticed she was off her feed and her yield had fallen drastically. Another cow, Iris, is walking lame, so David had to treat her as well this evening.

Graham was ditching most of the day since there is now little lambing. He does three feeding sorties to the sheep, down to Marsh, over the stream to Brinnen and up to Berry, and he reports they are all looking well with the early lambs already bold enough to break out on top of the banks.

Just one double born last thing this evening. There cannot be many more left to lamb now.

The following cows have been inseminated during this month:

Shamrock—Tuesday 8th

Iris—Tuesday 8th

An early primrose

BIRTH OF RAINBOW

This morning blue vast clarity of March sky
But a blustery violence of air, and a soaked overnight
Newpainted look to the world. The wind coming
Off the snowed moor in the South, razorish,
Heavy-bladed and head-cutting, off snow-powdered
ridges.
Flooded ruts shook. Hoof-puddles flashed. A daisy
Mud-plastered unmixed its head from the mud.
The black and white cow, on the highest crest of the
round ridge,
Stood under the end of a rainbow
Head down licking something, full in the painful wind
That the pouring haze of the rainbow ignored.
She was licking her gawky black calf
Collapsed wet-fresh from the womb, blinking his eyes
In the low morning dazzling washed sun.
Black, wet as a collie from a river, as she licked him,
Finding his smells, learning his particularity.
A flag of bloody tissue hung from her back end
Spreading and shining, pink-fleshed and raw, it flapped
and coiled
In the unsparing wind. She positioned herself, uneasy

As we approached, nervous small footwork
On the hoof-ploughed drowned sod of the ruined field.
She made uneasy low noises, and her calf too
With its staring whites, mooed the full clear calf-note
Pure as woodwind, and tried to get up,
Tried to get its cantilever front legs
In operation, lifted its shoulders, hoisted to its knees,
Then hoisted its back end and lurched forward
On its knees and crumpling ankles, sliding in the mud
And collapsing plastered. She went on licking it.
She started eating the banner of thin raw flesh that
Spinnakered from her rear. We left her to it.
Blobbed antiseptic onto the sodden blood-dangle
Of his muddy birth-cord, and left her
Inspecting the new smell. The whole South West
Was black as nightfall,
Trailing squall-smokes hung over the moor leaning
And whitening towards us, then the world blurred
And disappeared in forty-five degree hail
And a gate-jerking blast. We got to cover.
Left to God the calf and its mother.

MARCH

TUESDAY 1ST

There were bees out around the Dutch barn at Burrow. It is too early for them to be out, one late frost and they will never get home.

We cleaned out the tractor shed, filling the link box with the muck the tractors have brought in with them over the last year or so. The odd lost wrench turned up in the mud.

Still quiet on the lambing front with only one double and a single all day.

A few dry days now and we should be ploughing.

WRENCH
An adjustable spanner with grips on the jaws. A set of small tools is usually kept under the seat of the tractor. They are in daily use.

WEDNESDAY 2ND

The spring sun shone this morning. Graham saw butterflies up at Berry, and all the farm smells took on the mustiness of summer. Graham put all the gilts out in the orchard—three of them have not yet come into season and the warmth should help to bring it on.

We had to have the vet in for Emma who was panting hard at milking this morning. She had been reluctant to come out of her cubicle yesterday, but had milked well enough. Mr. Hindson confirmed David's suspicion—that she has contracted pneumonia. It is not bad, the injections should cure it. It is probably the sudden change of temperature over the last few days that has given her a chill—Channel Island cows are particularly susceptible to chills. We brought her in to milking this evening and she seemed quite happy—her appetite was good and her milk yield is down only a little.

I found one of the lambs down on the Marsh in trouble when I fed them at midday. It was a ram lamb, that could scarcely walk. His leg joints were badly swollen—so we may have to have the vet back tomorrow if he hasn't recovered.

An offer of 100 bales of good hay from Ted Hughes has boosted our stocks considerably. Graham, John and I picked it up from his farm over near Winkleigh this afternoon.

THURSDAY 3RD

Emma is a lot better this morning. Her breathing is back to normal and there is no temperature. But the lamb we picked up from the Marsh yesterday is not so good. The vet doesn't hold out much hope. The lamb is what is called "joint-ill".

The vet lanced the swellings to take off the fluid, but this evening the lamb is still looking poorly, although he is eating well enough.

Another dry, warm day and the primroses seem to have come up almost overnight. Burrow Lane is a mass of primroses and daffodils, and David is sure the grass has started to grow once again. If it goes on like this we will have enough hay to last us, and we will not have to buy in more.

The milk yield is down to 105—there are several cows drying off at present. But by the time the grass comes through and the herd is grazing outside, several more will have calved. Of course the grass will ensure a higher yield.

EWE HOGGETS
Last year's ewe lambs, lambing for the first time.

FRIDAY 4TH

Three doubles and a single this morning. Graham was kept busy. The ewe hoggets are lambing now. Five hoggets have lambed already, and some have produced doubles which is unusual for a hogget.

RINGING LAMBS
Neutering the ram lambs. This can be done either by ringing or by the knife. The ram lambs have to be neutered otherwise the flock would become impossible to manage.

Graham has begun ringing the ram lambs. It is clearly painful for them, but it passes quickly. They're running round again within 5 minutes. It takes time though. Some of the older lambs at Berry can already run fast so he has to drive them into a pen, separating the ewes from their

Hay for the sheep

lambs—and they do their best to escape.

Graham thinks we should be ploughing on Monday.

SATURDAY 5TH

David is still working away at the slurry pit. He has done more than a dozen loads, today and yesterday.

Harriet, a dry cow we expect to calve on March 11th, is looking less than her best. She's very large in the stomach but John thinks she has been thinning out too much recently. He suspects she could have twins. He put her out on Front Meadow with the in-lamb ewes this morning to give her some exercise and get the sun on her back.

Graham finished off the ringing this afternoon, and tomorrow he wants to get on with cutting their tails. There's a lot to do before the ploughing.

SUNDAY 6TH

We had a calf this morning but it wasn't Harriet's. One of the heifers calved down during the night. John suspected she might be close, but not that close. Pickright, she's called; and it's a Devon/Friesian bull calf, a big black one and as strong as you could wish for. The heifer is not big underneath—one of the reasons John thought the calving would be later than it was.

MONDAY 7TH

Harriet was delivered this morning of twins, one heifer and one bull calf. Both calves are big enough to be singles and Harriet is looking well. But it sounds better than it really is. Twins may seem fine, but in this case two are not necessarily better than one. Two bull calves would have been satisfactory, two heifers wonderful; but one of each is not so good. The bull calf is fine, but the heifer will be useless to keep on—we will never be able to breed from her. It is always the way, the heifer calf of mixed twins is infertile. Still, both are big enough calves and will do well in the market. Harriet is a good milker though, and with Pickright freshly calved they will inject some badly needed surplus milk into the herd.

Pickright knows how to kick. David discovered that this morning. Most heifers are nervous to start with but don't know how to kick accurately, but this one has the knack.

John has been cutting lambs' tails. Graham holds the lambs, John cuts. It is not easy to do or to watch, but it must be done. Come the summer and the warm weather, the flies would be at their dirty tails laying their eggs and giving them no peace.

TUESDAY 8TH

Dry enough for ploughing, and this afternoon in a gusty wind, Graham and John began ploughing up the ley on two of the Berry fields. They have been grass leys for 4 years and it was time for a change. There will be 22 acres of barley up there on Little Eastern Hill, Dutch Barn Field and Quarry Park and another 13 acres back at Parsonage. The ground turned easily and they ploughed until dark. We shall want to be tilling in early April.

John ploughs the erish

David took Rita's calf into market, but prices had dropped and he made only £27.

We moved some of the ewes into Long Close and fenced off a quarter of the root crop for them. There will be enough in that field for the flock until the grass comes through in a couple of weeks.

Pussywillows are showing their catkins along the hedges and the salmon are rising in warmer water down in the Torridge. Five have been caught in 2 days.

ROOT CROP
Long Close was divided in two. One half was drilled with kale, the other half with swedes for the sheep. The roots compensate the sheep when the grass is scarce.

TILLING
Strictly, the whole process of preparing the ground, sowing the seed and embedding it with harrow and roller. In common usage, however, tilling refers merely to the sowing of the seed.

LEY FIELD
A field already sown to grass after another crop.
A meadow is rarely ploughed up. Ploughing would destroy all the natural herbs that are so beneficial to stock. Front Meadow for instance has not been ploughed in living memory.

PLOUGHING
A long process even with modern machinery. With tractor and plough, it takes a day to plough 5 acres. With work horse and plough 5 acres would have taken 5 days.

DISC HARROWS
A set of circular steel discs that break down the furrows left by ploughing.

DRAGS
Heavy, spiked harrows. They are brought on after the disc harrows to break the soil down to a fine tilth, and to level the ground for tilling.

WEDNESDAY 9TH
On with the ploughing. A grey, windy day on top of Berry Hill for Graham and John, but good drying weather for ploughing. Ploughing up ley fields calls for precision. The plough must be biting down into 5 or 6 inches of soil, and there must be a constant well turned sod so that when it comes to harrowing and dragging the field later you don't pull the grass up through. On the other hand, if you plough too deep you bring up the subsoil and lose the value of the top 2 or 3 inches of well manured soil on the surface. The field has been used for sheep grazing for the last 4 years. Once ploughed we will have to take the disc harrows over it at least three times and then drag it—all this to break up the soil to a fine tilth. So there is a long way to go before we can till the barley.

David is still having problems with the new heifer, Pickright, down in the dairy. He can never keep on the clusters for more than a few minutes, and this evening she snapped the kicking chains. David's method of quietening her down was to pull up her tail and tie it to the rails. She calmed down after suffering this indignity.

THURSDAY 10TH
Graham went on ploughing on Berry Hill in gale force winds. John brought him up a hot dinner at midday, and by teatime this evening when he came back he had almost finished the Quarry Park Field.

Two more of the ewe hoggets lambed this afternoon, only singles but good lambs. The hoggets have been very promising. Most of them have lambed doubles and there are only a handful obviously not in lamb. Our rams are to be congratulated.

The milk yield went back over the 110 mark with Harriet already milking over 4 gallons a day.

FRIDAY 11TH
Graham did another morning's ploughing up at Berry and finished off Quarry Park. David busied himself with the

stock, feeding up everywhere, scraping down the yard and cutting kale.

John went off to a farm sale. He wants to buy perhaps six more cows this spring to bring the herd up to sufficient numbers to justify the investment in the new parlour. But there was nothing there of any interest.

The wind has dropped at last and the warmth is back in the air. You can smell the garlic already. It comes up in little green shoots everywhere under the trees. Once the chickens start pecking around in it we shall have the odd garlic egg, and it affects milk in the same way if the cows get at it.

SATURDAY 12TH

We never see Graham these days, he's been out all day ploughing Dutch Barn Field on Berry Hill.

Harriet is giving us cause for concern. The twins clearly left her weakened and she's been looking thin, but David saw there was something seriously wrong this morning. She seemed to leave all her cake in the trough, and after consulting John he sent for the vet. Mr. Byrne came in the early afternoon and diagnosed "Wooden Tongue". He injected her with an antibiotic. It is worrying because cows do not always respond to treatment with this condition. Some years ago there was an antibiotic that cured it but cows appear to be immune to it now to some extent. Mr. Byrne thinks she might pull through, but we can't be sure.

WOODEN TONGUE
An infection of the tongue which prevents the cow from swallowing.

SUNDAY 13TH

Another glorious day for ploughing, but it is Sunday, and on Sunday we do only what is essential.

Harriet looks to be making a good recovery in the shippen at Parsonage under John's tender care. He is feeding her extra oats to bring her back to fitness and she is eating well now.

The milk yield was up over 113 gallons this morning and that's an encouraging sign. That heifer, Pickright, was almost impossible to milk this evening. She is red and sore round her teats and kicked the cluster off a dozen times before we managed to milk her out.

MONDAY 14TH

It hailed and rained all night, so ploughing was out for the day.

Those Leicester sheep are getting out all over the place. They break down new fences and push their way through hedges. Graham is thinking of selling all of them soon.

We picked up a load of stones on Little Eastern Hill up on Berry this afternoon before we plough the field. It has always been a stony field, and the bigger ones interfere with the crop and need picking up regularly if we are ever to clear the field. It was windy up there in the hail and rain, but a rainbow curved over Iddesleigh towards the evening as the sun came through.

TUESDAY 15TH

Lucy, one of our best cows, and one we wanted a heifer from, calved during the night. David discovered them first thing this morning. She's delivered an average looking bull calf.

Appalling gales with lashing rain and hail made ploughing impossible again. Graham and David were fencing and banking most of the afternoon, interrupted temporarily by a stampede of Red Devon steers from a neighbouring farm which we had to herd back along the lane. Luckily all gates were shut, so there were no trampled gardens this time.

FENCING

The permanent fences are the banks. These are protected from livestock by a strand of barbed wire. Oak stakes are driven in at 5 yard intervals.

WEDNESDAY 16TH

Still too wet to plough and Graham is eager to get back up on Berry Hill. There is a field and a half up there still to do, Dutch Barn Field and Little Eastern Hill.

All our invalid animals seem to be improving. Emily and her companion are eating everything they're given, and they're both lively and alert. I haven't heard Emily cough for days now. Harriet is making a strong comeback. John is milking her out by hand each morning and evening, and she seems to look better every day. The lamb with the swollen legs seems more mobile now and the swelling is reduced, so he may be worth keeping on to fatten.

We are almost at the end of the kale now, and there are perhaps 2 weeks hay left. So it will be tight. If the weather warms up, and the grass comes through we shall be all right, but there is no possibility of putting cows out to grass in this wind and rain.

THURSDAY 17TH

This morning it was dry enough for Graham to set off up to Berry for the ploughing. John took up some hot pasties to him for lunch and they ate them out of the wind in the Dutch barn. Graham went banking after that and finished the 40 foot gap the lambs had made in the hedge, while John went on with the ploughing.

They've finished two of the fields now and begun Little Eastern Hill. While they were up there two more of the hoggets lambed—both singles. There must be less than a dozen left to lamb now.

One of our chickens is missing. There are a few holes in the

wire, so I think she must have gone out exploring for food and been unable to find her way back in for roosting at nightfall.

Parsonage

FRIDAY 18TH

No ploughing today—still heavy showers every hour or so and the ground never has a chance to dry. Instead Graham went hedging up on Berry and burnt up the hedge parings they cut last year—it was a fair blaze in spite of the rain. David was out spreading fertiliser on Upper Redlands. We spread this once a year on the grazing land, and dress the barley fields at the same time as we drill. Farmyard manure we spread every 3 years on every field.

SATURDAY 19TH

Rosebud calved—a few days early—and it is yet another bull calf, a good one.

Graham started drenching the ewes and lambs this morning—he started with the sheep out on Essythorn. They are drenched against stomach worms. It should be done when the lambs are about a month old, and Graham will be working his way through the flock over the next week or so, in between ploughing. He couldn't plough today—the rain was incessant.

David went hedging and fencing, mostly down at Burrow, before going off to Berry to fetch another load of hay for the cows. It is almost down to ground level up there—there may be another load or two there but certainly no more.

Graham drenching the lambs

SUNDAY 20TH

Harriet is still not as well as she should be. Graham thought about bringing her down with the herd, but John wants to keep her up at Parsonage for a few more days yet.

Graham got the cattle box ready for market tomorrow. We'll be taking in some sheep. The price of fat hoggets is high—about the only prices that are.

The clocks went forward last night. It has been getting lighter gradually every morning, but now we shall really notice the difference.

MONDAY 21ST

Fat market in Hatherleigh. There were just five ewe hoggets not in lamb and we decided to sell them fat. Graham took them into market along with the old sow that only farrowed five young last time. Graham had fattened her up over the last few weeks so she weighed over 22 score. She made £65, while the hoggets made £25 each.

One of the remaining hoggets lambed during the afternoon, a single.

David finds it takes him all the morning still to deal with the herd—milking, washing down, feeding the calves, feeding the cows, cutting the kale, feeding the kale, cleaning the yard and bedding the cubicles.

Graham went up to Berry in the afternoon to cut off overhanging limbs which would make headland ploughing awkward and shade the crops too much. Much of the wood is fit for pea sticks and lightings.

TUESDAY 22ND

There was a flood in the cattle sheds in the lower yard this morning and we had to call out Mr. Payne to mend a broken pipe—it seemed a good moment to clean out all the cattle sheds after that and David spread the muck on Furze Close.

Graham went back up to Berry to finish sawing the overhanging branches, and then went on with ploughing up Little Eastern Hill. David came up during the afternoon with the trailer to saw off more branches and load up with lightings.

WEDNESDAY 23RD

The ploughing up at Berry is finished this evening, all except for the hedge furrow. Graham will take up the single furrow plough to finish it all off.

David came back to Berry to pick up the last of the lightings and pea sticks. The pea sticks need tying in bundles and we will use them later on in the vegetable garden.

THURSDAY 24TH

Graham helped John cut lambs' tails—130 of them before lunch. They spent the afternoon ringing more ram lambs.

The cows were driven out on to Higher Redlands this morning. Fresh grass for the first time. There is not much of it, but the cows loved the freedom and the room.

BRUCELLOSIS

A form of contagious abortion. The germ attacks the womb of the cow and cuts the blood supply to the developing calf. A disease which has brought financial disaster to some dairy farmers.

The vet came to test the cows for Brucellosis—an obligatory Ministry test once every year to make sure there's no sign of the disease. We shall have the results next week. He also had a look at Harriet and gave her another antibiotic injection to stop her discharging.

David let the sheep out on to Haw Park to eat up the mixture of barley, grass and weed left over from last autumn's late ploughing. Haw Park was the field we failed to drill with grass seed before the ground became too wet last autumn. We shall plough it up again soon and try again this spring—in April most likely.

FRIDAY 25TH

There was no cockerel crowing this morning. Usually he crows at first light and again when we feed them after breakfast. There was an ominous silence this morning, and after

Crossing the brook

looking for the tell-tale flurry of feathers by the wire we discovered him lying dead and stiff under his perch. There was a speck of blood on his beak, but otherwise he was unmarked. It's a mystery—yesterday he was strutting around his kingdom as proud as ever. He was young too, barely a year old—a lot younger than all his hens.

Graham finished the headlands up at Berry, and David spread fertiliser on Front Meadow.

SATURDAY 26TH

The cows were kept in the yard this morning—there's no sense in poaching growing fields. The kale has all gone now and we are scratching around for the last of the hay. There is little doubt now that we will have to buy in at least 2 weeks' supply of hay to see us through. There is another bout of winter forecast, coming down from the north.

Graham brought the forty ewes and singles back from Berry to graze the root crop with the eighty that are already there. We want them to finish it up now. Then we'll be able to plough that field as soon as the weather improves again.

SUNDAY 27TH

Rainy, but the cows were given 2 hours out on Upper Redlands before David brought them back in for milking.

It feels as if we have gone back into winter. There is a frost forecast for tonight.

Cheering news from the dairy though. We are sending away 120 gallons a day now. Just those few hours of grazing each day has made all the difference to the yield.

MONDAY 28TH

The vet came. Lily's foot needed treating again. David has tried to treat it himself but she was limping this morning, and the vet has diagnosed a foot-rot condition and injected her as well as treating the foot itself. We are not allowed to send her milk away for 24 hours. Harriet is grazing in the long grass by the side of the farm drive and seems to have a good appetite, but she is still discharging from the infection she had after calving. Her Wooden Tongue condition is much improved.

Graham took the Tamworth gilt up to a neighbouring farm to be mated with the nearest Tamworth boar. Tamworth is a smaller slow growing breed and we are not hopeful of an economic return with the pig trade still deep in the doldrums. We shall probably sell her with her litter.

John and I went off to a dispersal sale. He is looking for just a few more cows to boost the spring calving, but prices were ridiculous—the good cows were fetching around £550 each.

RETURN

To come bulling. A cow returns every 3 weeks. There is excitement around a cow that is bulling. Very often she will be mounted by other cows. This is the most common sign that a cow is in season.

BARREN COW

Barren cows, and old cows that have outlived their usefulness, are usually sent to the fat market in Hatherleigh. There they are sold and slaughtered in the abattoir. Diseased animals, however, are not accepted for slaughter in the abattoir; they go to the knackers.

TUESDAY 29TH

David took in two bull calves to the market. They made £35 each.

David spend the afternoon burning up hedge-parings on Lower Redlands and Burrow Brimclose. The smoke drifted up almost vertically from the valley. Cold again but dry.

Another lamb today, but it looks a bit weak on its legs.

Dear old Hyacinth has returned again this evening, and it's for the third time. The oldest in the herd, and one of our two leaders.

Harriet has not improved. Although she is not discharging as much, she is looking thin and wretched. She may have to be sold.

WEDNESDAY 30TH

One of the last ewes, a two-tooth, lambed this afternoon. She had a poor lamb that has some difficulty in breathing. It sounds as if there is some fluid on the lung. The ewe is an attentive mother but the lamb hasn't the energy to get up to feed. Graham has injected it, but there's little hope.

A day of fencing, mostly around Essythorn and Long Close.

David sent away 123 gallons today. There is a noticeable improvement every day now. We are trying once more with old Hyacinth—no one is anxious to see her go.

THURSDAY 31ST

Graham took a sow down to Mr. Yelland's boar this morning. Mr. Yelland keeps a good Landrace boar so the offspring should be fair.

Everyone's sheep are breaking out, ours as well. There is not enough grass for them at this time of year, and once they've exhausted a field they look for a way out to greener pastures. So Graham moved eighty ewes with their lambs down to the Marsh where there is enough grass to keep them going. He is still feeding sheep nuts but it's the grass they need.

It is not the weather yet for either ploughing or drilling, but we will need to start drilling as soon as the weather improves. David and Graham got out the combine drill, took it to pieces, brushed off the cobwebs and rust and greased it up ready for the fine weather when it comes.

The following cows have been inseminated during this month:

Orchid—Thursday 24th
Violet—Tuesday 29th
Hyacinth 1st—Wednesday 30th

COMBINE DRILL
The machine that sows seed with the fertiliser all at the same time.

DRILLING
The method of sowing corn, using a mechanical drill.

HAPPY CALF

Mother is worried, her low, short moos
Question what's going on. But her calf
Is quite happy, resting on his elbows,
With his wrists folded under, and his precious hind legs
Brought up beside him, his little hooves
Of hardly-used yellow-soled black.
She looms up, to reassure him with heavy lickings.
He wishes she'd go away. He's meditating
Black as a mole and as velvety,
With a white face-mask, and a pink parting,
With black tear-patches, but long
Glamorous white eyelashes. A mild narrowing
Of his eyes, as he lies, testing each breath
For its peculiar flavour of being alive.
Such a pink muzzle, but a black dap
Where he just touched his mother's blackness
With a tentative sniff. He is all quiet
While his mother worries to and fro, grazes a little,
Then looks back, a shapely mass
Against the South sky and the low frieze of hills,
And moos questioning warning. He just stays,
Head slightly tilted, in the mild illness
Of being quite contented, and patient
With all the busyness inside him, the growing
Getting under way. The wind from the North
Marching the high silvery floor of clouds
Trembles the grass-stalks near him. His head wobbles
Infinitesimally in the pulse of his life.
A buttercup leans on his velvet hip.
He folds his head back little by breathed little
Till it rests on his shoulder, his nose on his ankle,
And he sleeps. Only his ears stay awake.

APRIL

FRIDAY 1ST

The lamb born on Tuesday died during the night and Graham has put in our only remaining orphan lamb with the ewe. He's had to tie the ewe up so that the lamb has a chance to suckle. It will be a relief not to have to make up any more bottles.

Graham and David finished fencing in Long Close where the sheep are now eating the last of the roots. Once they have finished we shall plough it up and drill barley.

SATURDAY 2ND

David spread fertiliser on Lawn Field and Burrow Meadow so the grass should come on this year. They are patchy fields both of them with overhanging oaks and elms preventing the grass from getting the sunshine it needs.

The bulk milk lorry is playing havoc with Burrow Lane, carving great furrows into the chippings. We shall have to concrete the lane at some stage, and at this rate of deterioration it will have to be soon.

SUNDAY 3RD

Graham cleaned out the pigs' house and the calves' house before lunch and fixed up the broken gate on Burrow Meadow.

A sunny Sunday with the ground drying fast all the time. It will be a week of harrowing and tilling if the weather stays right. There's a busy fortnight ahead.

HARROWING
The general term used for breaking down ploughed land.

John let the four big steers on to Ferny Piece to graze this afternoon and they are staying out tonight—the first night out after the winter.

MONDAY 4TH

John was thinking of taking two of the steers into the fat market, but we have been too busy.

Graham was up on Berry harrowing all day. He managed to finish two of the three fields. It's the first stage in breaking up the ground ready for tilling. He pulls the disc harrows along the furrows breaking up the turned earth, but not digging down far enough to disturb the turfs beneath. Freshly ploughed ley has to be disced at least two or three times—once with the line of the furrows and then once across them.

David stayed at Parsonage after he had finished the cows and fixed up the winnowing machine with John. Each year John keeps the best sample of his barley for the next year's seed corn. We need $2\frac{1}{2}$ bushels an acre and we're tilling just about 35 acres of barley, so there is a lot of winnowing to do. The barley has been stored all winter as it came off the fields last harvest, along with a certain amount of dust and mots and weeds.

WINNOWING MACHINE
A machine that separates the good large seed from the waste and small seed. It cleans the seed. Dust and waste are blown out behind. Small seeds are used for feeding the pigs.

BUSHEL
A measure of volume: 42 lb. of oats; 63 lb. of wheat; or 56 lb. of barley.

TUESDAY 5TH

Graham took two Friesian calves into market. They made £23 each, but he did not wait for the sale—there was the harrowing to do and the weather was right for it. He has harrowed both fields up at Berry and begun dragging them. He will need to drag them both again before tilling.

David took Harriet down to the dairy again this morning. She is still looking thin, but is a lot better in herself. He has to go down last thing at night to make sure she is lying in the cubicles out of the cold. She is not strong enough yet to spend as much time as the others outside so he brings her back into the yard for shelter after she has had an hour or so out in the fields.

The sheep are still breaking out in spite of the fencing. It is too cold for the grass to grow and they are impatient for it.

WEDNESDAY 6TH

Graham finished off the discing and dragging up on Berry. We shall have to buy some new discs soon. The set we keep up at Berry is perhaps 20 years old and the discs have been worn away. They should be the size of dinner plates—these look more like saucers. Then there was the ploughing of Haw Park to be started. Graham is still out there now in the dark.

David has been fencing all around to try to keep the sheep where they belong and then he winnowed several more bags of corn. He has assembled the drill so we could be tilling tomorrow now, if the weather holds fair.

Celandine, the kicking Guernsey, has rejoined the herd at Burrow. David still ties her tail up to stop her antics, and that seems to work.

John came back this evening after the sale with four new milking cows—Astor, Mayday, Alpine and Bernadette, all Friesians and all third calvers. Three have calved only recently and one is due in August. They averaged over £450. We have forty-one milking now and another seven dry up at Parsonage.

THURSDAY 7TH

Back in the depths of winter again with snow showers all day and a hard frost overnight. Any hope of the grass growing through has gone. It will be a long cold wait for spring.

The combine drill went wrong. John and David went to check it out while Graham ploughed up Haw Park. They tinkered with it for hours in the cold, trying to sort it out. By evening it was working as it should. No one can rest until the tilling is done now.

David sent away 156 gallons this morning. Of course that is a little more than it should be because the four new ones were not milked out on the morning of the day of sale. David is very pleased with them. They are gentle in the parlour and

A Hereford steer looking out

look to be fine prospects for the herd.

We had to borrow our first load of hay from Morley King down at Iddesleigh Mill Farm. Just eighteen bales, but we will need more later on if the cold stays with us.

FRIDAY 8TH

Good Friday and hot cross buns, but with the snow and frost in the early morning it seemed more like Christmas than Easter. There are more snow showers on the way, but meanwhile the ground is dry and right for tilling. John and Graham took packed lunches and two tractors off up to Berry and drilled all day. We don't like to work on Good Friday—but John will not rest now till it's done and we have to

CULTIVATING AND DRILLING

Like lambing and harvest, cultivating is a tense time for farmers. All the time there is the anxiety that the weather may prevent the next day's cultivating or drilling.

take the opportunity while we can. There may be snow around or rain, and he wants to get the corn in.

The yield is down a little to 154 gallons—that's close on 4 gallons a cow, but the cows are eating down what little grass there is all too quickly and there will be no more growth for a while.

David winnowed another 14 cwt. of barley this afternoon; there's still another ton to do.

SATURDAY 9TH

The drilling goes on in dry weather. Both Quarry Park and Little Eastern Hill are finished now—so there is only Dutch Barn Field over at Berry still to do.

While David was down at milking, Hettie noticed a ewe in trouble on Front Meadow. The ewe had lambed only the day before yesterday and was lying quite still. Hettie saw that she had a prolapse. She rang up the vet at once who came, pushed everything back in where it should be, and stitched the ewe up again.

PROLAPSE

The womb of the ewe falls and protrudes at the back. This happens only rarely, and usually in ewes about to produce a double. Sheep, like all animals, lie with their heads uphill at night. Gravity assists the prolapse which often occurs first thing in the morning.

SUNDAY 10TH

Easter Day, and with most of the drilling now done and the rain falling all day, everyone could relax.

MONDAY 11TH

It has dried off well during the night and John and Graham were able to get out early to finish up over at Berry. They tilled and harrowed the last field. Harrowing is the last part of the process and entails pulling light drags over the field to cover up the seed. One slight hitch, John ran out of seed corn, and had to run across to George Dunn's at Brimblecombe for a bucket and a half of corn for the last few square feet. But it's all done now, and although there is a lot more ploughing, drilling and harrowing to be done close to home, the main corn crop is in and there is warmer weather forecast.

We had to fetch another load of hay from Morley King's. We are not buying it directly—we will be letting him have some of our hay back in return when it's harvested. The grass still is not growing, although the rooks are nesting and the blossoms are out everywhere.

TUESDAY 12TH

Stock market. Graham took in the Leicesters, the white-faced ewes that seem to lead the break-outs. He took in two double couples and seven single couples. The double couples made £47 each and the single couples £37 a piece.

COUPLE

A ewe with a lamb is a single couple. A ewe with two lambs is a double couple.

We are filling in the ruts on Burrow Lane with stones, big stones to start with and then later smaller chippings to pack it down. David started on the worst ruts this afternoon while Graham spread manure on Furze Close.

The cows are lying out tonight for the first time this spring. The ground is dry enough, and although it is still cold at nights there's plenty of shelter in the lee side of the hedges on Burrow Brimclose. Anyway, there is hardly any straw left for their bedding.

Harriet is improving all the time now, there is a shine back on her coat; but Lily is still limping badly.

WEDNESDAY 13TH

John and Graham brought in the last of the loose hay from Berry this morning, while David spring-cleaned the parlour. He had to leave it unfinished; there was another load of hay to fetch from Morley King's farm.

Heather calved out on Front Meadow during the afternoon. David watched from the bank above just in case he was needed, but she calved down easily. Normally a cow will lie down to calve, but Heather got up when the time came and calved standing, so the calf had a slithery dive head first on to the ground. A Hereford cross Friesian heifer calf.

Graham spent most of the day ploughing up Long Close. John went in with the discs first to even out the tractor ruts we've been making all winter on the way to and from the kale.

THURSDAY 14TH

Long Close is finished, the headlands as well, and Graham has been ploughing up Haw Park. It turned easily enough. Up to now ploughing has been John's prerogative, but as the years go on he finds the cold more difficult to cope with. Graham seems as happy as John ever was to spend days on end on a tractor. He will finish Haw Park by this evening and then it will be just Little Rats Hill left to plough up for next winter's kale crop.

We had the vet in to look at Angelina who David suspected was still not in calf after four inseminations. Mr. Hindson confirmed this. So Angelina will have to be sold. It is a terrible waste, because she's a big, blackish Friesian cow of fine proportions—perhaps the best-looking one we have.

FRIDAY 15TH

The plough was going all day again and Little Rats Hill is all ploughed up except for the headlands. It was the first warm afternoon of the spring. The grass is growing through green at last and the horses are leaving any hay we offer them. The cows are out all the time now and need less and less hay. Correspondingly their yield is up—to over 160 gallons today.

LEAVING THE HAY

A sure sign that there is enough grass growing to satisfy the stock. Horses, cattle and sheep will always eat fresh grass in preference to hay.

There are fewer sheep breaking out now those white-faced Leicesters have gone; but then the grass is there to eat and there is no longer any need for them to go looking for it.

SATURDAY 16TH

Graham ploughed all day up on Little Rats Hill. By this evening he had finished that and all the hedge ploughing on Haw Park and Little Rats Hill. It was a fine day for ploughing with a dry breeze and warming sunshine.

David went fencing on Lawn Field during the afternoon. The grass is growing wonderfully well there now, but David wants to leave it another week before letting the cows in there.

I saw a lark rising over Upper Redlands this afternoon, the first of the spring for me. I lost him against the sun.

SUNDAY 17TH

Another summer's day. Still cold early, but the afternoon brought real warmth. There is dust underfoot again instead of mud.

MONDAY 18TH

Angelina was sold in the market today. She made £320—we paid £360 for her a year and a half ago. David took in two sows with her and they made £60 a piece.

Graham went discing on Long Close, but it came on to rain by mid-morning and he had to stop.

This rain is just what is wanted now. The first spring rain after the winter has dried out. We should start rolling the hay fields tomorrow—David went off to fetch the granite roller from a farmer who had borrowed it.

ROLLING

It is important to flatten the grass fields with a good rolling before the grass grows too long. The hay mower cuts quite low, so that any stones, molehills or lumps should be well rolled in. The granite roller weighs 1½ tons.

TUESDAY 19TH

The hoggets have started lambing again. We had almost forgotten there were a few still left to lamb. John and Graham had to call out the vet this afternoon for a small hogget trying to lamb what was obviously a huge lamb. Mr. Hindson managed to take it off, but it was dead. The hogget is recovering, but it's by no means certain she will pull through. Graham has to give her an injection of penicillin once a day. Three days should be enough to determine if she has the strength to survive.

David began rolling on Higher Redlands in ideal weather, the ground giving just enough to make a fine job of it.

Graham went down to the Marsh to bank up the hedge down there.

WEDNESDAY 20TH

Four more hoggets lambed today, two doubles and a single—all sound. The convalescent hogget seems well enough and is eating.

John put the younger cattle out to pasture this afternoon. This is the first time they have been out to grass and they were reluctant to come back in the evening. Bounce sorted them out. It is still not warm yet, but it will harden them if we let them out each day now for a few hours.

David recorded 160 gallons in the parlour and went on rolling Upper Redlands. Before milking this evening he spread fertiliser on Burrow Brimclose.

Graham went on with that hedge down on the Marsh. A solitary job down there all day but he is used to it. Most farming work is solitary.

THURSDAY 21ST

David finished Upper Redlands this morning and has made a start on Lower Redlands. The cows are still grazing on Lawn Field; there is more grass there than we thought. In the afternoon he hitched up the fertiliser spreader and drove over to Essythorn. It is poor grass there and needs a boost, so he's spread nitrogen which will bring the grass on. We don't like feeding too much nitrogen.

Graham was hedging and steeping all day down on the Marsh. There is still some way to go but he's gained several feet of useful grazing by doing it. He came back for tea but went straight out to disc Long Close until just before dark. Then the rain came down. Disced-up ground will hold the water much longer than ploughed—so drilling won't be possible there now for a few days even if the weather is right.

NITROGEN

Used in moderation it can do no harm, but continuous dressing of nitrogen could ruin a field. One of the great difficulties of farming is the compromise between the demands made upon the land so that it should be profitable and the need to maintain the fertility and balance in the soil. The maxim at Parsonage is that what is taken out of the soil must be put back.

FRIDAY 22ND

Two more hoggets lambed today, both singles. Just a few more left now; they should all be finished by the end of the month. The hogget that had trouble lambing a couple of days ago is looking more lively.

Graham was steeping down on the Marsh under grey skies, wrapped up well against a sharp gusty wind. David set up the winnowing machine again and managed to finish half a ton before milking time.

SATURDAY 23RD

Still cold and the grass has gone back a bit. The cows are eating it fast out on Lawn Field, and there is not enough warmth to bring it up again.

Back down in the parlour this evening, Celandine our kicking Guernsey was scouring and David drenched her against worms. Lily's foot is much improved though—she's limping still but it is barely noticeable. It's a lot messier now down in the parlour—the only disadvantage of grass grazing.

SUNDAY 24TH

The milk yield is down significantly. We have sun today, but there is a cold wind with it. We were hoping to save Lower Redlands for hay this year, but with the grass growing so slowly it looks as if we may have to move the herd in there. The grass on Lawn Field is all gone and David has had to move them on to Burrow Brimclose for the day. The cows must have continuous good grazing now if we're to get good milk yields. John is inclined to let them out on to Lower Redlands and if necessary buy the hay.

The wind did drop in the afternoon and brought an illusion of spring temporarily. I saw the first swallow down by the river; and basking on the nettles along Burrow Lane the brilliance of a Peacock butterfly.

MONDAY 25TH

Sunflower calved this morning—and this time we have got it right. It is a Hereford bull calf, tall and lanky with a black patch over each eye. Sunflower was a slow milker in her first lactation, but there's an enormous bag on her now. Let us hope she lets it down quicker this time.

Graham was able to disc Long Close this afternoon, the wind drying out the earth quicker than we expected.

John went to a sale near Crediton and bought three more cows—probably the last ones. The best, a first calver, cost over £400. The other two cows are second calvers.

TUESDAY 26TH

We sold Heather's heifer calf in the market this morning. A Hereford cross Friesian, she made £41—not too bad. Things seem to be looking up in the market. Assurances came also today for more money for milk producers. The three new cows milked well—there's 175 gallons in the bulk tank. David thinks we shall stabilise now at around 170 gallons, and with the grass beginning to come as it should and the need for large amounts of concentrates diminishing, there is room for a profit margin.

I called in the vet for Emily again. It's the same old problem—she must have caught a chill, but whatever it was she's off her food and wheezing away. Her temperature is up and Mr. Hindson injected her with an antibiotic and something to ease the breathing. As he says, the best thing she can do is to get out in the sun—but it's still too cold yet. At some

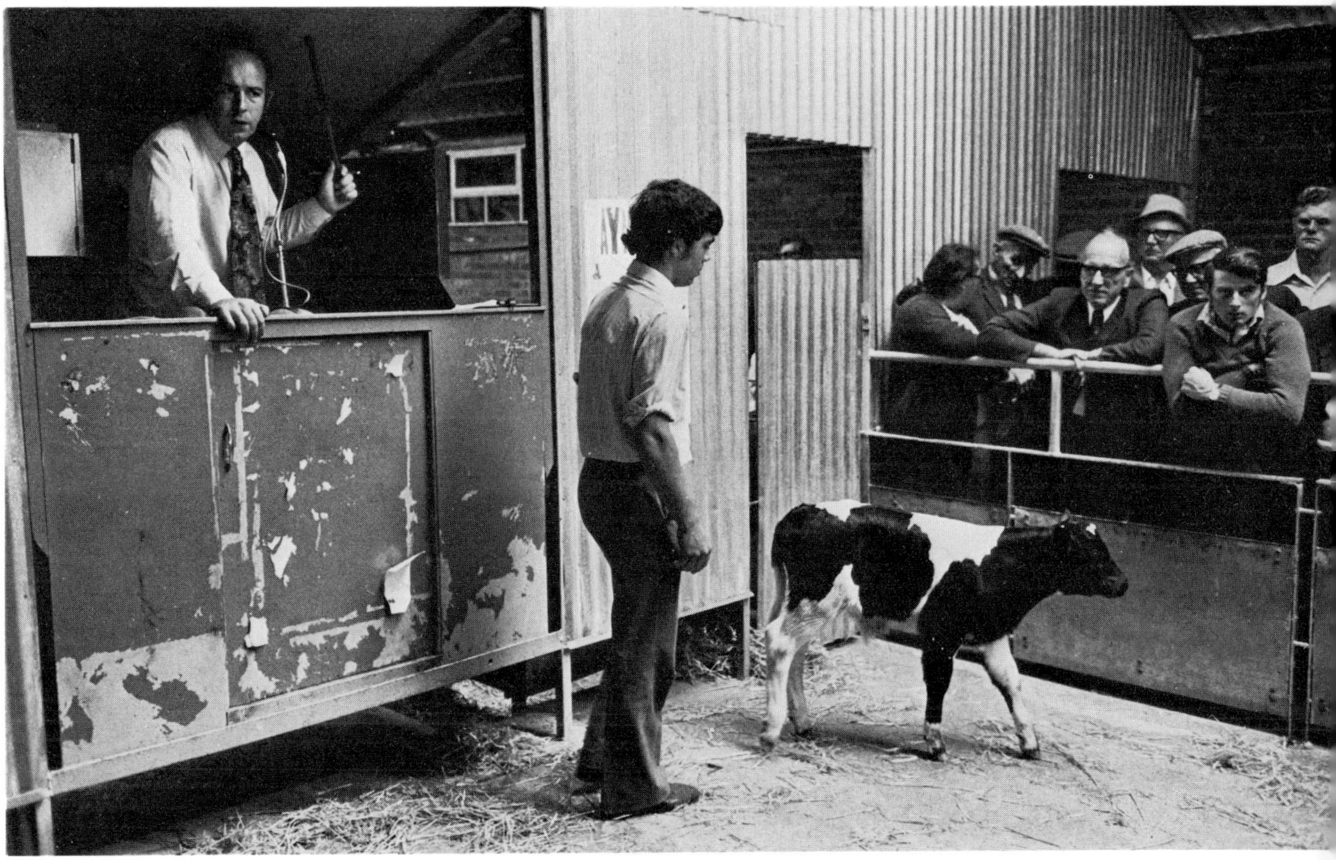

David selling a calf

stage soon she has got to stand on her own feet.

Graham cleaned down the dung spreader to its original luminous orange paint. We don't use it so much in the summer months.

WEDNESDAY 27TH

A warmer afternoon and we turned Emily and her Guernsey companion out to grass for the first time. The Guernsey reacted by frolicking around the field while Emily stood and contemplated a new world, coughing every few seconds. It is the first time either has been outside, and we brought them back in after a short time. Too much grass too quickly is not a good thing. Emily showed some interest in her nuts this evening, but has lost any real enthusiasm for food of any kind.

Graham and John started the docking after lunch. The clippers were greased up and the flock driven up into the yard. Fifty were docked before tea.

David took the loader down to Burrow and filled in the ruts in the lane with stones. But as fast as we fill them in, the milk lorry spills them out again.

DOCKING

The wool at the back of the sheep, around the tail, is always mucky. Each year at this time every sheep has the dirty wool shorn, or docked. The sheep stay cleaner in the summer and, as a result, more free from flies. This prevents maggot infestation.

Emily

DUTCH ELM DISEASE

There are over a hundred elms still living on the farm. It is unlikely any of them will survive the disease. Fortunately there are many oaks, beeches and limes, but the elms will be sorely missed. They are the biggest, most majestic trees on the farm.

THURSDAY 28TH

More rain and a cold wind blows with it. The elms along Burrow Lane have just sprung green, but we must watch them now for any sign of Dutch Elm disease. Last year we lost eleven trees. If the leaves turn yellow at the top of the tree or the bark looks as if it might be coming away, then we will know it is Dutch Elm disease. We would be very lucky to escape. In time they will all go.

John and Graham went on with the docking when it dried up in the afternoon. We've still a couple of fields to drill so we need a dry spell now.

FRIDAY 29TH

It must be the cold and damp that are affecting Emily, so we shall have to move her back up to a warmer house unless she improves. She is eating, but not enough to recover.

Too wet for docking, too wet for discing or drilling; so John and Graham began dismantling the old milk stand out in the lane, redundant now since we went over to bulk tank. There are some useful sleepers and blocks to bring back. Like most farmers, we are great hoarders, believing in the principle that most things will come in useful if you hang on to them long enough. Much of what is "used" we can re-sell, baler cord, paper bags, etc. and the rest always seems to find a use. The sleepers we shall probably use for covering the well in Lower Redlands.

Some of these new cows are milking wonderfully well—Mayday milked over 8½ gallons today, with the froth on top she almost fills the jar on the morning milking—over 5 gallons.

SATURDAY 30TH

We moved Emily to a warmer house and fed her some fresh grass. She ignores it at present, preferring her dry hay, and it is the dry food that aggravates her lung condition. Still, she's chewing her cud again this evening, and looks more composed.

John and I made a start on clearing out the Lower yard and one of the cattle houses down there. We shifted a mountain of dung, before the backs began to ache. Graham drove in more ewes for docking—he finished thirty more before evening.

The following cows have been inseminated during this month:

Emma—Thursday 7th
Poogly—Sunday 10th
Lavender—Wednesday 20th

STEALING A TROUT ON A MAY MORNING

I park the car half in the ditch and switch off and sit.
The hot astonishment of my engine's arrival
Sinks through the 5 a.m. silence and frost.
At the end of a long gash
An atrocity through the lace of first light
I sit with the reeking instrument.
I am on delicate business.
I want the steel to be cold instantly
And myself secreted three fields away
And the farms, back under their blankets, supposing a
plane passed.

Because this is no wilderness you can just rip into.
Every leaf is plump and well-married,
Every grain of soil of known lineage, well-connected.
And the gardens are like brides fallen asleep
Before their weddings have properly begun.
The orchards are the hushed maids, fresh from
convent—
It is too hushed, something improper is going to happen.
It is too ghostly proper, all sorts of liveried listenings
Tiptoe along the lanes and peer over hedges . . .

I listen for the eyes jerked open on pillows,
Their dreams washed with sudden ugly petroleum.
They need only look out at a sheep.
Every sheep within two miles
Is nailing me accurately down
With its hellishly-shaven starved priest expression.

I emerge. The air, after all, has forgotten everything.
The sugared spindles and wings of grass
Are etched on great goblets. A pigeon falls into space.
The earth is coming quietly and darkly up from a great
depth,
Still under the surface. I am unknown,
But nothing is surprised. The tarmac of the road
Is velvet with sleep, the hills are out cold.
A new earth still in its wrapper

Of gauze and cellophane,
The frost from the storage still on its edges,
My privilege to poke and sniff.
The sheep are not much more than the primroses.
And the river there, amazed with itself,
Flexing and trying its lights
And unused fish, that are rising
And sinking for the sheer novelty
As the sun melts the hill's spine and the spilled light
Flows through their gills . . .

My mind sinks, rising and sinking.
And the opening arms of the sky forget me
Into the buried tunnel of hazels. There
My boot dangles down, till a thing black and sudden
Savages it, and the river is heaping under,
Alive and malevolent,
A coiling glider of shock, the space-black
Draining off the night-moor, under the hazels—
But I drop and stand square in it, against it,
Then it is river again, washing its soul,
Its stones, its weeds, its fish, its gravels
And the rooty mouths of the hazels clear
Of the discolourings bled in
Off ploughlands and lanes . . .

At first I can hardly look at it—
The riding tables, the corrugated
Shanty roofs tightening
To braids, boilings where boulders throw up
Gestures of explosion, black splitting everywhere
To drowning skirts of whiteness, a slither of mirrors
Under the wading hazels. Here it is shallow,
Ropes my knees, lobbing fake boomerangs,
A drowned woman loving each ankle,
But I'm heavier and I wade with them upstream,
Flashing my blue minnow
Up the open throats of water
And across through the side of the rush

Of alligator escaping along there
Under the beards of the hazels, and I slice
The wild nape-hair off the bald bulges,
Till the tightrope of my first footholds
Tangles away downstream
And my bootsoles move as to magnets.

Soon I deepen. And now I meet the piling mob
Of voices and hurriers coming towards me
And tumbling past me. I press through a panic—
This headlong army of river is a rout
Of tumbrils and gun-carriages, rags and metal,
All the funeral woe-drag of some overnight disaster
Mixed with planets, electrical storms and darkness
On a mapless moorland of granite,
Trailing past me with all its frights, its eyes
With what they have seen and still see,
They drag the flag off my head, a dark insistence
Tearing the spirits from my mind's edge and from
under . . .

To yank me clear takes the sudden strong spine
Of one of the river's real members—
Thoroughly made of dew, lightning and granite
Very slowly over four years. A trout a foot long,
Lifting its head in a shawl of water,
Fins banked stiff like a schooner
It forces the final curve wide, getting
A long look at me. So much for the horror:
It has changed places.
Now I am a man in a painting
(Under the mangy, stuffed head of a fox)
Painted about 1905
Where the river steams and the frost relaxes
On the pear-blossoms. The brassy wood-pigeons
Bubble their colourful voices, and the sun
Rises upon a world well-tried and old.

MAY

SUNDAY 1ST

There was December-white frost first thing this morning, but by the time milking was finished we were back in May with the sun already warm. The cows were waiting by the gate and jostled to come into the parlour for their cake—no reluctant milkers this morning.

Aptly, Mayday—currently our best milker—was bulling this morning. But it is only the first time after calving, so we will wait till she comes round next time.

There is a sow due to farrow in a few days. Her udder has sprung, the skin stretched taut, the teats pointing out.

MONDAY 2ND

Graham finished docking the last sixty ewes.

The sow due to farrow is still due. The nest is all chewed up and ruffled, shaped like a boat, just as it should be but there's little sign of her settling to farrow.

John decided this was the time to drill Long Close with barley. It is half-past nine now and dark, but John is still out there. The drilling is all done and he's harrowing it over. We need the rain now—both John and the weather forecasters say we are going to get it.

Emily is better, on a diet of freshly picked grass from the banks on Front Meadow she seems to be fighting back. In spite of the cold the grass is really growing now—Upper Redlands looks like the beginning of a hay field, and the barley up on Berry is already shooting up through.

One of the fat steers—a massive Friesian slipped his chain this evening and helped himself to an entire bag of cattle feed. It's serious—cattle can die just as easily from over-eating as from hunger. He is looking very dejected and is clearly suffering. It may pass through the system, in which case he will scour out. If he does, all will be well.

TUESDAY 3RD

Spring show market in Hatherleigh, and we took in weaners and couples. The weaners were not good enough to show, but with pigs so costly to fatten we decided to sell them out as weaners. There is just a hint of optimism about the pig market. Our five weaners made £20 a piece, more than we expected. It could be that there is still some hope for pigs.

Graham took in five white-faced Leicester single couples, and took second prize for the Leicester class, selling the couples for £42. He is pleased with his second prize, and we have now sold most of those Leicesters that led the break-outs earlier in the spring.

There were show classes for everything in the market, calves, steers, pigs, sheep; and the place was full. It was raining of course and that means there is less to be doing on the farms. I have rarely seen Hatherleigh market so crowded. All the poultry cages are filling up—there were ducklings, chicks, sitting eggs, goslings, cockerels—and a few rabbits as well.

SHOWS
In Hatherleigh market there are several show sales a year, that is a combined show and sale when breeders compete with each other to produce the best stock. Everything is for sale by auction as usual. So that like should compete against like, the show is divided into classes of breeds and trade requirements.

It rained hard most of the morning, so the barley John drilled yesterday had a good start. We were going to drill Haw Park with barley, but it will take a few days now to dry out and it is getting a bit late. Probably it will go to grass as we originally intended.

The sow has milk in her udder—we know that much, Graham tested her this evening. It will not be long now.

WEDNESDAY 4TH
The steer that helped himself to the cattle feed is going to have to be killed. The vet came this evening and diagnosed ruminal acidosis, a form of alcoholic poisoning. The barley is fermenting inside him, disrupting the digestive process, and there is little chance he can pull through and remain the same weight. He was due to be sold fat next Monday anyway. John drenched him with Epsom salts to help him during the night, and tomorrow first thing Graham will take him off to the abattoir. We hope he won't die during the night. The vet is fairly sure he won't, but if he did we would lose over £300 on him. A dead carcass may be worth £10; to slaughter he could make above £300.

POISONING
Always a danger with stock. Two main causes are poisonous plants in the fields and banks, such as woody nightshade or ragwort; and the chemical poisoning taken in when the stock lick lead paint from walls, or eat pastures too recently sprayed. If the carcass is contaminated there is no question of it being used for human consumption. All meat is rigorously inspected at the abattoir.

The sow still hasn't farrowed by this evening, but Graham is sure it will be tonight. She escaped from the orchard during the evening and wandered off—a sign she will be farrowing any time.

THURSDAY 5TH
The sow farrowed during the night, after her little walk-about yesterday. She had only seven little pigs, and that is a disappointment. However, the sow looks well, she's got milk and the little pigs are feeding.

The Friesian steer went away this morning, and Graham took a Hereford steer with him. The Friesian may not be fat enough to be first grade, which means he will fetch less money per cwt., but the Hereford was top grade and will do well. We shan't know how much they make until we get the weights and grades through from the abattoir.

Graham went dung spreading across the stream on Brinnen. It is the last of the dung spreading, it is getting too late really. We like to do it early spring and autumn if we can. He cleaned out the spreader some time ago, but the wet, cold weather has delayed the summer grass. There's no harm in spreading manure so long as enough rain falls to wash it in.

The grass still is not coming as it should. The milk yield is down, it has been down by a few gallons each day. It is down to 175 gallons—it was 190 at one point last week.

FRIDAY 6TH

More driving rain and cold. Getting up for milking was difficult this morning, and the cows looked dejected as they came in. Dairymaid stood on my hand and Buttercup whisked me round the face with a filthy tail.

Emily is doing well on her hand-picked grass—I wish it would warm up and dry up—then she could go out and get it

Devon and Hereford steers

herself. She eats everything we have time to pick. Her coat is shining again and she skipped this morning when she saw the wheelbarrow of grass.

The Government announced a small price rise on milk for farmers. We shall need it.

A cuckoo called out across the valley this afternoon, informing us it was spring. It is the first I've heard; and indeed he must be right because in spite of the wet and the cold, the hedges are dotted all along with bluebells and campion and violets. A canopy of leaves—beech and elm—is beginning to cover Burrow Lane.

MILK PRICES
A rise of about 4 pence a gallon to a new price of 45 pence a gallon. The price paid to the farmer fluctuates with the grass season. Farmers are paid less in the summer and more in the winter. Although any price rise will help the small dairy farmer to be profitable, the cost of cake and hay affects profitability as well. Like all aspects of mixed farming, the farmer expects fluctuation in market price and costs.

CUCKOO
Usually first heard in mid-April in this part of Devon.

SATURDAY 7TH
The little pigs are doing well. They sleep and twitch most of the time under the lamp. They are of uniform size and already plumping out.

The hay situation is desperate. Everyone is scratching around for the last bales of hay from the depths of the barns. No one has any left to sell. By now there should be no need for it. The grass should be growing through, but it isn't.

Cold and dreary. I went off to Bratton Clovelly this morning to buy another cockerel—a Maran. He's fine and tall with flowing tail feathers. We shall have to cull some of our older Rhode Islands this winter and we will need young hens to replace them.

HAY
It is always difficult to estimate hay consumption in advance. Although consumption per head of cattle can be calculated, it is not possible to allow for a long winter particularly after a below average hay harvest.

SUNDAY 8TH
The new cockerel crowed exultantly for most of the day. The only happy sound on yet another grim grey day.

If this grass doesn't grow soon, there will be a late hay crop, and little of it. To remain at all self-sufficient in hay we must take 130 bales off every acre. The milk yield is down again.

MONDAY 9TH
Graham took in three fat pigs to the butcher this morning, but with a £3.50 subsidy on each pig there was some paper work to be done before he could deliver them. It was to the market first for grading and registration, then back to the butcher for killing. We shan't know for a day or two exactly how much we will be paid for them. The pigs are fewer in the sty now, but with sows being slaughtered all over the country the market must surely turn soon.

There's a heifer due to calve within the week now—we are not sure of precisely when because she was bought in-calf having run with a bull elsewhere, but by the look of her udder John thinks it won't be long.

PAPERWORK IN THE MARKET
A licence is required to move pigs into the market. Then a Ministry certificate of grading has to be granted.

TUESDAY 10TH

David took in Sunflower's calf to the store market this morning. Bidding was brisk. David had expected £65 for the calf—a good Hereford cross Friesian bull calf. The bidding started at that figure and went on up to £79—one of the best prices of the day. It would have been good to keep a calf of that calibre, but £79 now with no costs to set against it, is difficult to resist.

Wet again, so we spring-cleaned the parlour in the afternoon, and cleared out the sides of the Big Barn.

WEDNESDAY 11TH

Lamb drenching kept us busy all day. The lambs should be drenched every month before weaning against stomach worms. All the lambs—over 200—were done by this evening. It's a back-breaking job—you have got to catch them, hook the drench into the mouth, squirt it down and then mark the ones you've done.

But the day was not over. This evening Tulip 2nd calved, with John and Graham pulling away at the calving ropes. It's a fine Friesian heifer calf. She was easy and quite quick for a heifer, but she found it difficult to settle down. She kept lying down with her back legs up against the wall. Then she would have to be stood up again and again until she lay down finally in the centre of the old barn to calve. The calf stood up within 15 minutes.

Warmer again this evening and the swallows were flying low at dusk. That could mean rain tomorrow.

THURSDAY 12TH

A day spent clearing up for the summer months. We scooped out most of the calves' houses, the shippen and the yard and Graham loaded it all up to be dumped in the orchard. It is too late now for dung spreading, so we shall spread this lot in the autumn. It is all looking very clean and unfamiliar in the shippens.

Tulip 2nd was taken down to the milking parlour with her calf this morning. David put the calf in the back of the pick-up and the heifer trotted along after. It was her first initiation into the big herd and the routine of the milking parlour, and David left her until last so that he could take his time with her. She milked down a gallon and never kicked at all. This evening she was a bit livelier, but compared to some she behaved like a veteran.

In spite of more rain and a cold wind, the barley on Berry is up 3 or 4 inches already.

FRIDAY 13TH

The day tried its hardest to live up to its reputation. One of

Fetching the cows for morning milking

the little pigs born last week was found dead this morning. It was the smallest and had a distended stomach—Graham thinks it was having difficulty in passing food through its system.

David noticed Poogly was off colour at milking this morning. She gave less than a gallon and looked wretched. But it was not until after milking that he found her lying down in the yard outside unable to get up and shaking from end to end. Staggers was his first thought. He rang up Mr. Hindson, who came out within five minutes—the symptoms demanded speed. Staggers kills in an hour if it is not caught in time. Luckily the prognosis was not too serious. It was an acute attack of colic, brought on by her coming into season again. She had been inseminated last month but apparently

STAGGERS
Hypomagnesia, a drastic drop in magnesium content in the blood causing spasms. Often fatal if not treated quickly.

COLIC
Severe indigestion, leading to inflammation of the large intestine.

unsuccessfully. He injected her and said she would be up and well by the evening. And so it proved, but she did not come in season, at least not noticeably; so insemination will have to wait another 3 weeks.

The cows are milking well again. They've been out on Lawn Field over the last couple of days and the yield is back up over 185 gallons again.

SATURDAY 14TH

Today is Club Day in Iddesleigh. Most of the men living around the village are members—and it is an exclusively male concern. In the morning they parade the banner through the village behind the Hatherleigh Silver Band and march up to church for a service. There's a drink at the pub after church, and then the Club dinner served on the long tables in the village hall.

There were sports in the afternoon and the pub stayed open—so it was mostly children's sports.

David and Graham worked during the afternoon. Graham took advantage of the dry weather to disc over Haw Park. But in the evening they both came up for the social in the hall.

IDDESLEIGH

The nearest village, a mile away, with some 250 people on the electoral roll. A hundred years ago there were twice the number of people living in the parish. The drift away from the land was caused by ever increasing farm mechanisation. There are two churches, one Anglican and a Methodist chapel; a public house (the Duke of York) a post office and a small shop. Most people work on the land or in the allied industries such as meat marketing, forestry, etc.

CLUB DAY

The annual celebration of the Iddesleigh Friendly Society founded in 1837 to provide some insurance for farmworkers' families against illness and death. It was an early form of agricultural trades unionism. Friendly Societies existed in many villages in this part of Devon, but now there are only two left.

Club Day celebrations in Iddesleigh

LEATHER-JACKETS
The larvae of the cranefly or daddy-long-legs. They are slug-like bugs that eat corn. Serious damage to the crop can be caused if they are not eliminated quickly.

SUNDAY 15TH

Leather-jackets have been found in the barley up on Berry. John spotted a poor patch on Dutch Barn Field, and sprayed it immediately. The spray along with a good rolling tomorrow should see them off.

There was a frost last night again, and I noticed that the ash trees are still mere skeletons. They seem to know when it is wise to break bud. They are the last to come green in the spring, and the first to turn in the autumn. "Oak before ash, we only have a splash. Ash before the oak and we shall have a soak."

MONDAY 16TH

Buying and selling. Four more fat pigs went to the butcher this morning. He will tell us the dead weight later, but all four were good and fat.

Mr. Goss over at Barwick was selling his small herd this afternoon, and we all went over. John took a fancy to one cow, due to calve in 3 weeks' time. He paid over £400 for her, but the calf should bring that back to £350, all being well.

David and Celandine will have to part company soon. We have tried using that Guernsey as a nurse cow, we have tried milking her with the herd and she still kicks anything or anyone that comes near her. She has battered the calves, and this morning kicked David in the chest. He has used kicking chains on her, tied up her tail, but it is no use. We have heard there are kicking bars that prevent cows from lifting their legs to kick. We shall have to get some.

TUESDAY 17TH

Last week's heifer calf went to market this morning and made £35. The market was down a fair bit, but she was a heifer and that makes all the difference. Over the year we want to sell our calves at an average of £50 a head: that pays for the water, the electricity and the vet's fees.

One calf gone and another arrived. Sorrel calved around lunch time. She was mooching round Front Meadow looking restless, so John brought her in away from the others and helped her calve. She was bought last year in calf to a Hereford bull, and today she has had a strapping bull calf, even bigger than the one we sold last week. Sorrel is a good milker, heavy and quick; next time we shall try to have a Friesian heifer calf from her.

With the ground drying off, David rolled out Lawn Field and the drier parts of Burrow Meadow. This eradicates all the hoof marks and rolls in any loose stones.

Graham and John harrowed Haw Park, working it up for drilling tomorrow, if the weather holds.

WEDNESDAY 18TH

Warm and dry enough to drill barley on Haw Park. It is late, later than John has ever drilled by a full fortnight, but then everything is at least 2 weeks behind in terms of growth. However John reckons it will have a good chance. He drilled the barley first thing this morning and by lunch had harrowed it in. Then in the afternoon he sowed grass seed over the top of it, sowing it with the fertiliser spreader. A final harrow and it was done. If all goes well, there could be a light grain crop at harvest, and a grass crop coming up underneath that we shall be able to use for sheep grazing in the autumn.

We had the returns from the abattoir where we took the two steers on 5th May. Both made over £330.

UNDERSOWING

Sowing grass seeds immediately after drilling the barley, so that after the barley is harvested the grass is ready to grow on.

THURSDAY 19TH

Today you could smell the warmth in the ground, and the grass is really beginning to grow back. Suddenly there is plenty to eat everywhere and we have stopped worrying about hay. Hay harvest must be late this year. Usually we expect it as May turns into June but it will be mid-June or later this year. The grass on Upper Redlands is up to a foot high, and Front Meadow is not far behind.

Emily is poor again—she refuses fresh grass and her concentrates. We will give her a chance outdoors tomorrow, but we are not optimistic.

WILD FLOWERS

Found in great profusion along the banks, in the hedges and in the fields. The high banks represent a sizeable acreage, being often 4 feet high on either side and 4 feet wide at the top. Much of the land has been and is permanent pasture for sheep, and this has meant that the flower population could grow undisturbed by the plough. There are also areas of uncultivated land where wild flowers abound. Farmers spray less intensively than in many parts of the country, and this must be helpful.

Commonly seen in the hedges are Blackthorn, Dogrose and Honeysuckle; and on the banks Primroses, Bluebells, Campion, Stitchwort and Tufted Vetch. Some wild flowers rare elsewhere are often found here. Among these are Bastard Balm, Scurvy Grass, Pennywort and Hartstongue Fern.

FRIDAY 20TH

A strong warm breeze from the east and there is very little mud about any more. Even the muddiest of gateways has turned to dust. Front Meadow is full of buttercups, and bluebells and cow-parsley pattern the hedgerows.

We tried Emily outside again today. She's been off her food and languishing for 2 or 3 days. Yesterday she ate nothing. I watched her out in Lawn Field and I don't think she ate anything all day. When I drove her in this evening, she had to stop every few yards to fight for breath. It's the finish.

SATURDAY 21ST

We called in the knackers to destroy Emily this morning. They put her down with a humane-killer gun and took away the carcass. It has been a long struggle, and perhaps many farmers would not have begun it. But where there's life we have to try. It was doubtful from the beginning whether she would make it. But there comes a time when the animal has

HUMANE-KILLER

A gun with a with-held bolt. The gun is placed on the animal's forehead above the eyes so the bolt enters the brain. Death is instantaneous.

to take over and support itself. She would never have made a cow unless she had that strength. We shall miss her. It is sad to have lost.

There is a dairy inspection from the Ministry of Agriculture on Monday, so we had a thorough spring-clean down in the parlour this afternoon.

Hyacinth 3rd, a heifer in Front Meadow will be calving any day now.

SUNDAY 22ND

Hyacinth 3rd has looked uncomfortable now for 24 hours but she has still not calved. I had to go up close to check her—the grass has grown so long in Front Meadow that the calf might have been hidden.

The barley drilled 3 weeks ago on Long Close is coming up wonderfully well—the entire field is green again.

Hyacinth 1st was bulling again this evening—the fourth time she's returned. She will be sold fat at the end of this lactation. She has been a fine cow, but is well past her best. She is a devil to milk. Her udder hangs so low that it is difficult to put the cluster on at all. The one-time leader of the herd, she has been deposed now by Beth and by her own daughter Hyacinth 2nd.

MONDAY 23RD

The man from the Ministry of Agriculture came to inspect our dairy and passed it with enthusiasm. He reckoned it to be one of the best-kept dairies on his patch, so we are all glowing with pride. This evening the cows made more mess than they have ever done, and there was a power cut in the middle of milking so we didn't finish till 10 o'clock.

Graham spent most of the day on Little Rats Hill, discing and fertilising. We shall be putting in more kale so we won't run out early whatever the weather next year.

The hay fields are already high. A week of sun has turned everything round in the right direction.

TUESDAY 24TH

Hyacinth 3rd calved early this morning. It's a Devon cross Friesian heifer calf, jet black and shiny. We brought her in under the lamp to warm her through but she's weakened gradually during the morning. The vet was called in to cut the umbilical cord which appeared to have ruptured in the calf causing a nasty swelling. She improved after this, and is now just managing to suckle. She's still weak but the chances of survival now are good. Fortunately, the mother is calm and will stand still for suckling. But her teats are far too wide apart at the back, hanging almost outside the line of the back legs, so she won't be easy to milk.

Graham and Bounce rounded up the sheep this afternoon, caught the lame lambs and trimmed up their feet. All that wet weather has brought on foot trouble. We had to treat about a dozen. Bounce is a much improved dog, responding quickly and surely now to Graham's whistle and using his intelligence more. He seems to be losing the impetuosity of his early days.

WEDNESDAY 25TH

Hyacinth 3rd's calf is suckling on her own now and seems to improve with every hour. She looks strong on her legs and there is no longer any need for the lamp.

I spotted a deer down by the brook when I was feeding the lambs up on Brinnen this morning. It moved away quickly into the trees but there's no doubt it was a deer—I saw the back legs clearly enough. John last saw one on the farm 10 years ago.

Lizards were out basking on the hot tarmac along the lane.

Hyacinth 2nd calved this evening—a nice Friesian heifer calf. Our calving luck is turning.

THURSDAY 26TH

Kale drilling. After working down Little Rats Hill, John sprayed it with a pre-emergent spray, harrowed it in and then drilled the kale in the afternoon and finished just before sundown. He drilled in two directions to ensure thicker covering over the 5 acres.

Harriet is looking poorly. She walks slowly, measuring every step and seems to hang her head. She milks well enough in spite of it, but we won't be keeping her long now.

David went shooting up on Berry this evening after milking. The rabbits have been getting at the young corn and have damaged it badly in places, particularly near the woods. He saw several but never got close enough to shoot.

Graham moved the sheep on to Lawn Field, to get the grass down for Jubilee. The village is having its Jubilee Sports out on Lawn Field, so we will let the sheep in for a couple of days, then roll it and mow it.

RABBITS

A menace to young corn. Although still affected by myxomatosis, they still breed in sufficient numbers to pose a threat. Shooting is one way of limiting the rabbit population.

FRIDAY 27TH

Drenching the milking herd. It took most of the morning to drench the cows. They were all awkward to drive into the crush and fought against the bars while they were inside. We were drenching against fluke and worms. It was an upsetting morning for them. The disturbance and the drench itself will mean a drop in the yield over the next 48 hours and a messy dairy. They always scour afterwards.

FLUKE AND WORMS

Roundworm, tapeworms and fluke reach the pasture as eggs in the droppings. The eggs hatch out into larvae and are eaten by the grazing animal. Worms thrive on damp pastures, another reason fields should be well drained.

Hyacinth 3rd's calf has definitely pulled through. She is thinner than she should be, but feeds strongly now.

SATURDAY 28TH

Some farmers are shearing already and if this weather goes on we shall be shearing next week. We could be shearing now but the combs and cutters are away being sharpened and they won't be back for a day or two. Still, there is the shearing shed to be set up and cleaned out, and the pens to be fixed up. This is hot weather for the sheep; they don't move much unless they have to. They graze in the cool of the day and rest up in whatever shade they can find.

Meanwhile we filled in the ruts in Burrow Lane and did some hedging on Burrow Brimclose.

This last week of sun and warmth has changed the landscape. Everything is growing through green. Even the walnut tree by Burrow Lane has decided to come out now. Quite suddenly there are trees between us and Dartmoor—and the perspectives have altered.

COMBS AND CUTTERS
The comb is the stationary bottom edge of the clipper and the cutter moves from side to side across it cutting the wool.

TREES
This part of Devon is covered in small woods and spinneys. The trees are of great importance to the farmer and to the land. They provide shelter for the stock against sun and rain. They provide a haven for wild life, especially for birds. But pasture and crops growing in the shade of trees are of poor quality and slow to grow.

SUNDAY 29TH

The milking machine hiccuped this morning—the vacuum feeder on one side has failed. So milking was a long process—just four cows at a time. The engineer will be here tomorrow.

We picked out five lambs fat enough to grade—that's 90 lb. live weight—for tomorrow's market. They will be our first to go—later than usual.

MONDAY 30TH

The fat lambs made £24 each—not a wonderful price. Three fat pigs made £35 a piece, and Poogly's Aberdeen Angus cross heifer of 18 months ago failed to grade and made only £160. Jersey cross calves just will not grow away. A dismal market for us.

The milk yield is down, and well down. The drench and the continuing dry weather are to blame. A spot of rain last night, but a blazing hot afternoon again.

I saw a sparrowhawk swoop on a wagtail in the lane—an efficient kill—a flurry in the dust.

One of the tractors was serviced this morning. We have better brakes now and a bright new exhaust to replace the old rusty stack.

TUESDAY 31ST

A better day in the market. Graham took in two calves, Sorrel's 2-week-old Hereford cross Friesian (£79) and Hyacinth 3rd's Devon cross Friesian heifer (£55). So the calf

David turning the feed selector in the parlour

market is back up to its earlier levels.

It has been a day spent setting things up. We put the lighter wheels on to the tractor and got the sprayer ready for tomorrow. The corn fields need spraying with weedkiller. There are thistles and nettles and all sorts coming up and we should spray now while it is dry. It was too windy this afternoon, but we are all ready now for tomorrow. We like to do it on a still evening when all the bees have gone.

David dug up the horse thistles on Brinnen. They grow to over 3 feet and after a campaign waged over some years there are only 100 of them left. We do it once a year to keep them at bay, and each year there are less to dig up.

The kale is already peeking through on Little Rats Hill. It is always reassuring when you see a crop come through for the first time.

The following cows have been inseminated during this month:

Alpine—Friday 6th
Bernadette—Tuesday 10th

SPRAYING
It is important that the spray lands only where it was intended. Spray will drift for half a mile in high winds and damage susceptible crops.

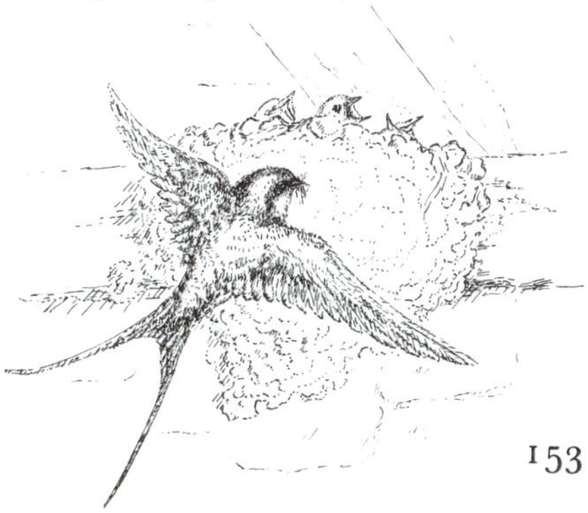

LAST LOAD

Baled hay out in a field
Five miles from home. Barometer falling.
A muffler of still cloud padding the stillness.
The day after day of blue scorch up to yesterday,
The heavens of dazzling iron, that seemed unalterable,
Hard now to remember.

Now, tractor bounding along lanes, among echoes,
The trailer bouncing, all its iron shouting
Under sag-heavy leaves
That seem ready to drip with stillness.
Cheek in the air alert for the first speck.

You feel sure the rain's already started—
But for the tractor's din you'd hear it hushing
In all the leaves. But still not one drop
On your face or arm. You can't believe it.
Then hoicking bales, as if at a contest. Leaping
On and off the tractor as at a rodeo.

Hurling the bales higher. The loader on top
Dodging like a monkey. The fifth layer full
Then a teetering sixth. Then for a seventh
A row down the middle. And if a bale topples
You feel you've lost those seconds forever.
Then roping it all tight, like a hard loaf.

Then fast as you dare, watching the sky
And watching the load, and feeling the air darken
With wet electricity,
The load foaming through leaves, and wallowing
Like a tug-boat meeting the open sea—

The tractor's front wheels rearing up, as you race,
And pawing the air. Then all hands
Pitching the bales off, in under a roof,
Anyhow, then back for the last load.

And now as you dash through the green light
You see between dark trees
On all the little emerald hills
The desperate loading, under the blue cloud.

Your sweat tracks through your dust, your shirt flaps
        chill,
And bales multiply out of each other
All down the shorn field ahead.
The faster you fling them up, the more there are of
        them—
Till suddenly the field's grey empty. It's finished.

And a tobacco reek breaks in your nostrils
As the rain begins
Softly and vertically silver, the whole sky softly
Falling into the stubble all round you

The trees shake out their masses, joyful,
Drinking the downpour.
The hills pearled, the whole distance drinking
And the earth-smell warm and thick as smoke

And you go, and over the whole land
Like singing heard over evening water
The tall loads are swaying towards their barns
Down the deep lanes.

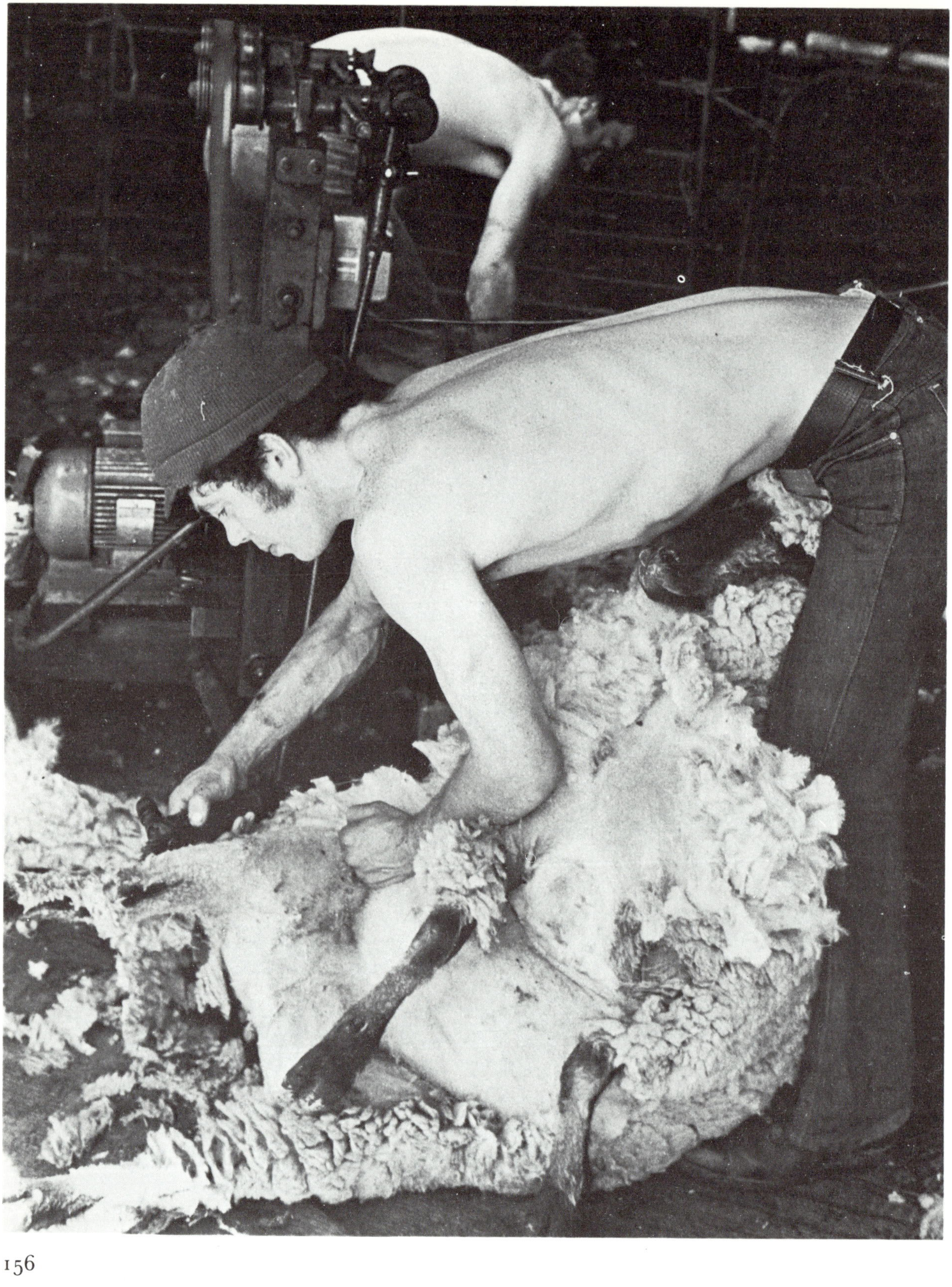

JUNE

WEDNESDAY 1ST

A day of uninterrupted problems. From the moment the vacuum in the milking parlour began fading at morning milking to when a wild heifer stamped on John's foot.

The parlour is functioning, but only just. By this evening the feeble suction on the clusters was irritating the cows and they played up accordingly.

We have got the new kicking bar to stop the kicking cows in the parlour. It clips on over the rump and under the flank above the udder, and it works well. Bluebell was upset by it and panicked this evening, trying to turn around and escape from the parlour. At one moment it looked as if she might break through the bars.

KICKING BAR

A great improvement on the kicking chain, the bar prevents the cow from lifting her rear leg to kick.

THURSDAY 2ND

The milking machine has been mended. Graham and I milked the herd and even Bluebell behaved like an angel. We still haven't sprayed the weed in the barley fields; the wind was too high again today.

FRIDAY 3RD

The wind dropped sufficiently for John to drive over to Berry and spray two fields of barley with weedkiller.

There's another heifer, Shamrock 2nd, looking as if she might calve any day now. We are bringing her into the barn at night so we can keep an eye on her.

SATURDAY 4TH

Graham began the shearing this afternoon. The combs and cutters came back razor sharp yesterday. Graham and David shear differently, but basically they shear the fleece into two halves, starting with the neck and legs, and then under the stomach, before removing the main fleece. A nick here and there is inevitable—the sheep never keep still for long. The fleece is rolled up, tied around and then stored in the corner of the barn. An average fleece may be worth about £2.50 this year; and it all goes to the Wool Marketing Board at North Tawton where it is graded for quality. We have over 350 to shear, including the lambs. But we start with the adult sheep. We could have shearers in to do the work, but two

Opposite:
Sheep shearing

together can shear ten in an hour and anyway it's cheaper to do it ourselves. Graham started late today and was working on his own. He managed fifteen by milking time.

SUNDAY 5TH

Heavy rain fell during the night and by this morning the mowing grass was lying flattened in the fields. It is not spoiled though, the sun will bring it up again.

The cows came in crisply black and white to milking this morning, with mud on their feet.

Graham spent much of the afternoon out cutting weed on Lawn Field. We are holding the Jubilee sports there on Tuesday and it needs trimming up a bit. We expect about 200 people from the parish.

WEED-CUTTING

Thistles, nettles and docks are the most common weeds, and must be cut regularly if they are to be contained. The finger mower is used on the flat fields where the tractor can go, but on the steeper slopes the weed has to be cut by hand.

MONDAY 6TH

A day preparing for Jubilee tomorrow. Mowing and rolling the Lawn Field, putting up flags and bunting and tidying around.

This evening there was a bonfire on a hill above the village and Morris dancing around the green. The rain came driving in from Dartmoor, but the fire, piled high with tractor tyres, blazed on in spite of it. On the hills all round we could see the fires of Monkokehampton, Exbourne and Okehampton.

TUESDAY 7TH

Jubilee day began with grey skies and sheeting rain, but by the time we went to church at 2 o'clock the glass was going up and we held our sports in warm sunshine.

Somewhere in between tennis and football the milking was done and stock fed, but this was no day for farming and the cows were hurried through their routine and turned out smartly.

It will be good to get back to normal tomorrow. The cows will appreciate it too.

WEDNESDAY 8TH

In between the showers there was time enough to finish spraying the weeds on the barley in Long Close. The barley is already tall and green; it has shot up during this spell of wet weather. John's toe is back now to its normal size so he finds the tractor work a lot easier.

One of the gilts farrowed in the sty just after lunch. She has farrowed eight little pigs and they all look healthy enough. Perhaps by the time they are fit to sell, the pig market will have improved.

THURSDAY 9TH

Shamrock 2nd, the heifer that looked fidgety and ready to calve some days ago, finally calved this afternoon, by herself and out in the Front Meadow. She has produced a Hereford cross bull calf that is already steady on his legs by this evening. There is another cow ready to calve in a day or so—the one John brought back from the local sale 3 weeks back.

With eight cows going dry now, all of them due to calve in the autumn, the milk yield is down to 160 gallons or thereabouts. Harriet is still giving a fair drop of milk, but her walking is more and more unsteady and she may have to go before she finishes her lactation.

Too cold and wet for shearing. The fleece must be dry and the weather warm so that the sheep do not catch a chill.

FRIDAY 10TH

Another grey day with violent rainstorms and thunder. Work outside was kept to the minimum until after lunch when the cloudburst abated and we could get out around to see to the sheep and lambs.

Graham moved the dry cows down to the Marsh where there is a carpet of lush grass, and brought the three nearest to calving back up to Furze Close. There are two to calve now within the week and one more soon after. But the only warm and dry place to be was the pigs' house, so he cleaned those out before going off down to milking this evening.

The mowing grass is lying flattened in the fields. We shall need a week of sun to put it right.

MOWING GRASS

The name given to grass intended to be cut as hay.

SATURDAY 11TH

Sleepy this morning—the milking machine refused to suck and Graham and I twiddled all the knobs we knew until we discovered we had forgotten one in the engine room. The cows waited patiently enough but we didn't get our breakfast till late this morning.

Out feeding the sheep I found one on her back up against the hedge over on Brinnen. I turned her over and she galloped off in panic at her rescue. It is one of the commonest ways that sheep die. It usually happens when they are in full fleece and in the peak of condition. They roll over to scratch themselves and are incapable of righting themselves—not unlike a beetle or a tortoise. A few hours like that and they die, or the rooks come and peck out their eyes.

Graham deloused the new litter of pigs. This seems a strong healthy litter with a good mother.

DELOUSING

A routine measure to protect the litter against lice which always seem to lodge themselves behind the ears of the new-born pigs.

SUNDAY 12TH

I found a tawny owlet up against the stone bank on Burrow Lane just before milking this morning. It was camouflaged against the stone and it was only because one eye blinked black and shiny that I picked it out. It's not injured, just exhausted and cold at spending the night out of the nest. Back home in front of the stove it revived and this evening it is sitting up and eating a dinner of minced beef and duck's feathers. An owl shuts his eyes when he eats—a kind of supercilious ecstasy.

STONE BANK
Many banks, particularly near the houses and along roads, are faced with stones to prevent the banks from falling out. The stones also provide some protection for the bank against stock. Laying stone banks or stone ditching is not done now very often because it is very costly.

Lucille, the last cow John bought, calved in the big barn this afternoon. She took her time, seemed uncomfortable and unsettled so John and I took off the calf, a large bull calf, a Friesian. He will make good money in the market, but we need heifers.

The sun shone for the first time for a week, and this evening the mowing grass has sprung back up again on Upper Redlands. We could do with 3 weeks' sun now to get the shearing done and the hay in.

MONDAY 13TH

Graham took eighteen fat lambs into market this morning. They made £24.50 each. As more lambs come fat we shall take more in each week from now on. These ones were all born back at the beginning of February. We're feeding them on oats and lamb pellets to fatten them more quickly. John keeps saying we must cut down on the ewes, that we haven't got the grass, but he's been saying that for a year or two now.

LAMB PELLETS
Lambs are fattened quickly. They are fed a mixture of oats and lamb pellets during the suckling period so that they put on flesh quickly. Lamb pellets are a compound with a barley base.

Whilst Graham was gone, Marigold calved down a day or two early in the big barn. She calved quickly and easily, a Hereford cross heifer calf. So yesterday it was a bull when it should have been a heifer and today it was the other way round. Still, the calf is living and Marigold looks to be in good condition.

David has been unwell for a day or two and has been off work, but he's a lot better now and will be back in harness soon.

TUESDAY 14TH

Graham cut weed for most of the morning. Weed seems to grow whatever the weather. There were a lot of thistles down on the Marsh when we drove two more dry cows down to join the others.

The corn is a foot high now up on Berry Hill and an even growth; so the rabbits and the leather-jackets didn't eat much. And everywhere the mowing grass is long enough to cut if only it would dry up and allow us to get on with it.

WEDNESDAY 15TH

It is dry, but still not warm enough to promise any hay harvesting in the next few days. But we want to be ready when the time comes, so David serviced the mower. John has already bought the baler cord. The baler itself was checked over this afternoon, parts replaced where necessary and greased up. Just the hay turner and the elevator to see to and we shall be ready.

The milk yield is around 160 and rising, and with the machine working well now, and David back; life is easier for everyone. From tomorrow we shall be able to send away both Marigold and Lucille's milk and that will boost the yield. The last of the dry cows, due to calve in August has gone up to Parsonage, so we should hover around the 160–170 gallon mark for most of the summer.

THURSDAY 16TH

Another day waiting for the warm weather. Graham went weed-cutting down on the Marsh around the hedges.

We drove the cows up to Furze Close after milking this evening. The grass is not thick on Lower Redlands or Burrow Brimclose—it is just not growing fast enough to keep

MOWER

Operated by power take-off from the tractor, our mower has two rotating discs that cut the grass. To prevent dust and flying stone, the mower is surrounded by a large rubber skirt like a hovercraft. It cuts close to within an inch of the ground.

DRY COWS

For the last 8 weeks before calving, the cow is left unmilked or "dry". This is a gradual process with the amount of cake lessened slowly. These 8 weeks give the cow the chance to build up strength for the calf inside her as well as a full supply of milk. The milking cow's routine is as follows:

60 days after calving she is inseminated.
305 days after calving she is dry.
2 months after going dry, she calves.

Checking the mower

pace with the cows' demands. There is maybe a week's good grazing on Furze Close so that should give some time for the Burrow fields to recover.

It warmed up towards evening, and we managed to shear a few sheep after feeding the pigs. We had the flock in anyway to drench the lambs for the third time.

FRIDAY 17TH

Another calving. Clover calved down just before midday. John bought her last year in milk and in calf to a Hereford bull. All went well and she gave us a sturdy bull calf. She has sprung a huge udder so she should help to boost the milk yield a little.

With the weather dry but still October-cold, it was another day for weed-cutting. Today David and Graham were out with hooks and scythes and cutting down the docks and nettles and thistles on steep ground where it is too dangerous to take the tractor. If we left them to grow they would seed and make matters worse.

There's just a chance we may go on shearing tomorrow if it stays dry tonight.

SATURDAY 18TH

John sprayed Haw Park this morning—the barley field undersown with grass. It had to be done with the narrow tyres on the tractor so as not to damage the growing barley. We shall need to put the wide tyres back on now for haymaking. Some of the hay fields are not flat; narrow tyres would be dangerous on any kind of slope.

TRACTOR TYRES

Many tractor tyres contain water as ballast, to ensure stability and maximum purchase power on the tread. They last for about 5 years and cost about £60 each on the big tractor.

More shearing this afternoon. We drove sixty more ewes into the Big Barn, and with both David and Graham on the clippers they managed to shear thirty before tea. The fleeces were dry all through and came away easily. Any dampness and the likelihood is that the fleeces will not grade; that would mean money lost on every fleece. Some of them struggle a bit at first but a judicial knee on the neck and a firm hold is usually enough to quieten them down.

SUNDAY 19TH

With twelve cows dry—one in four of the herd—the milk yield is down to 155, but it is unlikely to fall any further. On average one in every six cows in a herd is dry. We've hit a low patch with a heavy summer calving ahead. The Marsh is the most level field we have and all twelve are down there together. It is important that cows heavy in calf are kept off the slopes.

Meanwhile the herd is still grazing Furze Close for another day or so. That means a trip up and down Burrow

Lane. They've rubbed away at the banks in the lane to such an extent that Graham has had to put up hurdles to stop them from pulling the bank away.

The owlet we took in has grown feathers where there was down, and picks up food himself now. He can fly across the shippen, but always on a downward trajectory.

OWL
Tawny owls are the most common owls in this part of the country, though the sight of a white barn owl flying low along the hedges at dusk is not unusual.
Among other birds seen around the farm are buzzards, herons, jays, rooks, crows, plovers, kingfishers and the more common garden birds.

MONDAY 20TH

Not enough fat lambs to make it worth a trip to market, and anyway the lamb market has slumped somewhat so it will not hurt to wait a week.

It was hazy all day, with the glass going up steadily; John went out to look at the mowing grass at 8 o'clock this evening. The signs were all good and by 9.30 he had cut the first five swaths of Higher Redlands. Haymaking has begun, and it looks like a good crop off this field, maybe 1,000 bales off the 7 acres. The quality is better than we expected with more leaf lower down the stalks. Now we need uninterrupted dry weather for 10 days or so. Ideally we should finish cutting this field tomorrow, let it wilt a little, turn it twice on Wednesday and Thursday and bale it on Thursday evening.

SWATH
One line of cut mowing grass. The mower cuts its first swath around the outside of the field and then works inwards.

We shall try to finish one field at a time—the weather is too unsettled at the moment to risk cutting all the hay fields.

TUESDAY 21ST

Higher Redlands is cut and has been drying for the afternoon under a hazy, reluctant sun. John finished it off in the morning, and he is fairly optimistic that it will be dry again tomorrow and ready for the first turning.

Graham took the two Hereford calves into market—one bull, one heifer, Shamrock 2nd's and Marigold's. The bull made £76 and the heifer £45. He was going to take in Lucille's bull calf, but he needs to fill out a bit. Marigold saw her calf in the link box ready for market when she came up to Furze Close after milking, and was most upset, turning every few steps and advancing on me menacingly. But she relented and merged back into the herd.

Graham was shearing on his own and managed another twenty sheep this afternoon. He has done most of the shearing and so far his back is holding out. The fleeces are all thick and of good quality.

The longest day of the year—it is still light at 10 o'clock this evening.

WEDNESDAY 22ND

A blazing day with intense, still heat in the afternoon. Sheep shearing was exhausting work, but by this evening there were only a few ewes left to do. It takes about 5 minutes

actually to shear a sheep, 5 more for catching, tying up and marking. It's a smelly, sticky affair, with the lanolin from the sheep's wool greasing hands and arms and trousers. As we tire the sheep seem to struggle more and each one takes that much longer.

John turned all the hay in Higher Redlands so that it is drying now on the green side and lifted so the air and sun can permeate through. John stayed out turning and came back late for lunch.

The bullocks up on Berry broke out last evening, and David went off up there to bring them back. They had knocked down a fence and clambered through into the mowing grass, knocking it about and no doubt enjoying the temporary paradise of long fresh grass. There's not too much damage done.

HAY
One of the two harvests in the summer months. Barley follows in August. Although quantity is important, quality is just as vital. Continuous fine weather is necessary so that the hay can be made quickly. The quicker it is made, the better the quality.

THURSDAY 23RD
The last of the ewes were shorn this morning in heavy heat. There are the lambs still to do in July, but all the ewes are done now and they stand thinner and whiter in the fields. So the clippers are stored away for a few weeks. Whilst we had the flock up in the yard, Graham picked out the fat lambs for market next Monday. He weighed them and found about twenty heavy enough to go at 90 lb. live weight. They lose 2 lb. on the way to market and once killed their live weight is halved. All the sheep are over on Furze Close now, picking where the cows left off. There is plenty of grass there to bring more lambs up to the 90 lb. mark before Monday.

Opposite:
John cuts the mowing grass for hay

An old granite cross by the River Torridge, thought to be a prayer stop for monks

With the weather forecast looking promising John decided to risk cutting the mowing grass on Great Eastern Hill at Berry. There is 12 acres up there and he cut it all before teatime. The mower came back covered in a yellow carpet of pollen.

Graham turned the hay again on Higher Redlands ready for troning and baling tomorrow, if we get the sun. It is still very wet all through. Every morning there is heavy mist and dew and it takes several hours of sun to dry it out.

TURNING, TRONING AND BALING
Once cut, the hay is turned at least twice. The turned hay is then troned, or gathered into lines ready for the baler.

The milk yield is up again in the dairy. The cows are grazing Lower Redlands where there's a rich crop of grass and the selling of the two calves on Tuesday means that all the milk from those two cows is now going into the tank. A calf will drink 2 gallons a day when it is sucking the cow.

FRIDAY 24TH

By 10 o'clock this morning when the early cloud and mist was still hanging over us we knew the weather had broken. The day stayed dry until a fine drizzle set in for the evening, but it meant we could not bale the hay lying on Upper Redlands. Before baling it needed 2 or 3 hours of sun, but it never came. The cut grass up on Great Eastern Hill is still green, so delay and rain at this stage will not be too harmful; but with the dried hay on Higher Redlands which was almost ready to bale, the rain is a set-back. Every hour it stays out there in the wet and damp reduces the quality of the hay we harvest. A prolonged spell of wet weather with no opportunity to turn or bale and the crop could be spoiled. So John is cloud watching, looking for signs. This morning, the cattle were carrying their tails high, stampeding in pain from warble fly bites; and the rooks were cawing raucously on the wing—both signs of more rain.

David cut weed and thistles down on the Marsh, leaving the cut grass for the dry cows to eat up behind him. He cut just a dozen swaths leaving enough for the cows to clear up before he cuts again.

SATURDAY 25TH

Rain and drizzle, but the glass was going up. A day spent washing down the tractors, cleaning down the yard and fencing down Burrow Lane.

Bernadette has developed mastitis, the first case we have had for some time. Her udder was very hard in one quarter, and David treated her with antibiotic. The milk will have to go to the pigs.

The monthly milk cheque came in with the morning's post. It works out that there's a gross profit on the month's milking of around £500. A few months like that before next winter and we shall be happy.

SUNDAY 26TH

The decision to make the hay on Upper Redlands was reached by mid-morning. The sun was shining through light cloud and there was a drying breeze. Graham turned it once more before lunch and then we began work. John set off with the baler and by half-past seven this evening he had finished baling—1,485 bales. David and I stood the bales up in stooks, four in a stook, so that they could dry out before loading.

Haymaking has little of the romantic associations that people imagine. There is the continuous rhythm of the baler picking up, baling and disgorging; the dust dries the throat and the string on the bales tears at the fingers.

We expected only 150 an acre off that 7 acre field, so 200 bales an acre is good going. But quantity is not everything. That day out in the rain has taken some of the sweetness from the smell of the hay. It is good enough, but not excellent.

We worked on until 10 o'clock, and managed to bring in four trailer-loads before it became too dark. There are ten

A farming decision

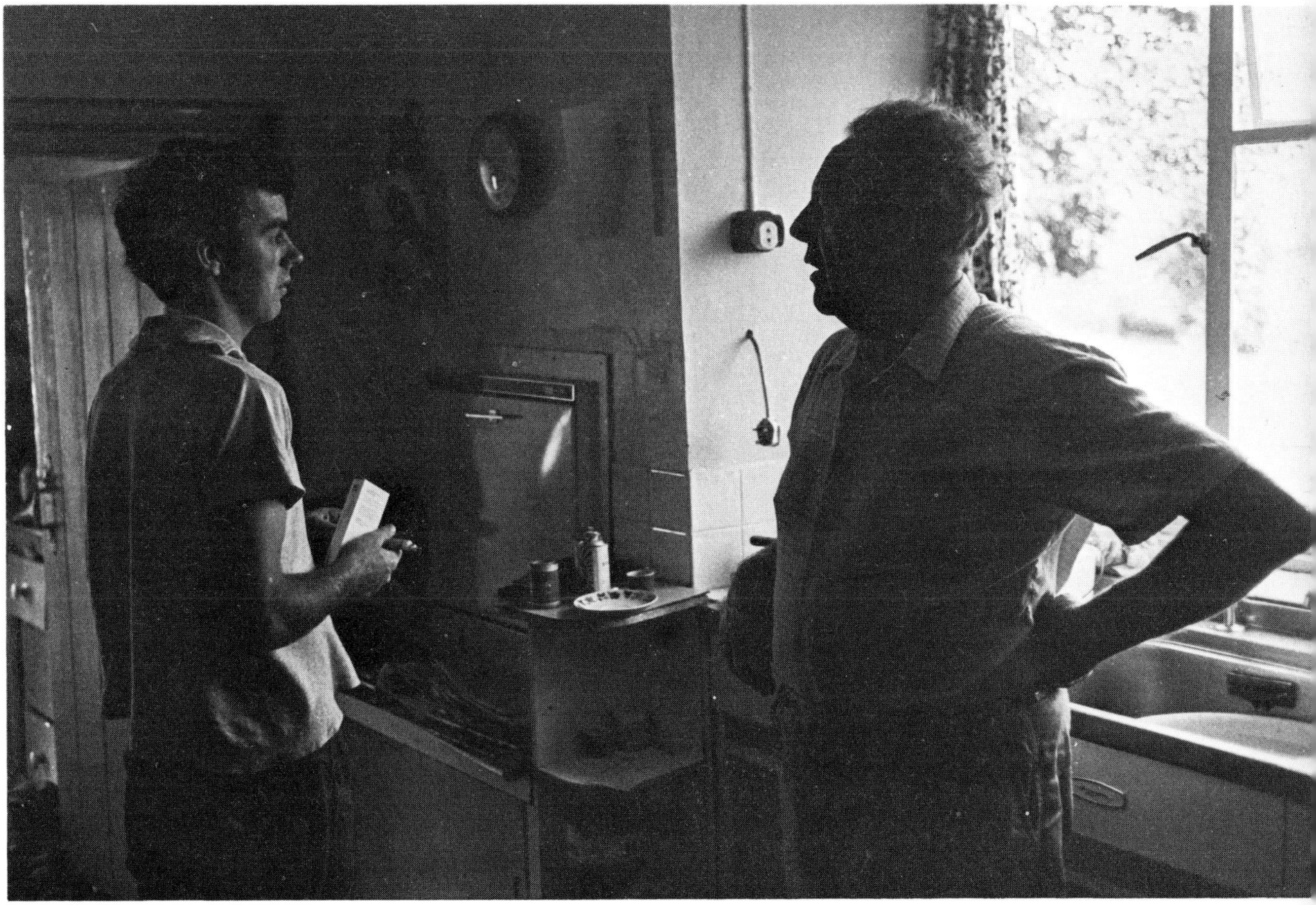

loads still left out there. We should have it all in by lunchtime tomorrow if it stays dry.

MONDAY 27TH

It threatened to rain all day, but John ignored the threat and the haymaking went on. All the bales from Higher Redlands are in under cover in the Dutch barn down at Burrow. It's a green-looking crop but at least it is all in safely. Graham went up to Great Eastern Hill to turn the hay up there while John stayed behind and cut Front Meadow. The hay off Front Meadow is usually the best on the farm, full of the natural herbs only found on unploughed pasture. It makes top quality hay most years. This year it may be a little short—we have had stock in there from time to time since the spring because of the shortage of grass.

There were lambs ready for market, but no time to take them. It is hay before everything at the moment. John is living on his tractor.

TUESDAY 28TH

Good prices in the store market. Lucille's Friesian bull calf, and Friesians are much favoured on the continent, made £87, our top price of the year; and Clover's Hereford cross £67.

A savage downpour mid-morning put paid to any hopes of haymaking. With all 12 acres on Great Eastern Hill cut and turned and lying out, and a third of Front Meadow cut, we must have a few warm days. Of course, the weather is a constant concern all around the year, but at haytime it becomes more acute because rather than worrying about a general build up of dry or wet weather and its consequences, we're looking only to the next afternoon, or the next morning to see if the hay can be harvested.

To keep us occupied while waiting for the rain to move away, there were the pigs' houses to be cleaned out.

WEDNESDAY 29TH

Still no chance of getting the hay in. It was dry, but there was no warmth in the air. John thought of turning in the afternoon, but a walk out on to Front Meadow was enough to put him off. It was far too damp for turning.

John and Hettie settled down to the accounts, paying bills and checking figures. The accounts are done purely for the accountant and the tax man. At any one time John knows just how unprofitable or profitable is any one of the farming operations. At present, for instance, the cows are doing us proud, the lamb market has taken a dip, calves are selling well, and the less said about the pigs the better. John keeps a tight watch on all the markets and costs, and makes a farming decision to increase or decrease investment or production without reference to monthly accounts. It may sound strange, but on a mixed farm where you are dealing with fluctuating markets, to rely on what the books say can be treacherous. It needs a farmer's instinct and years of experience, and even then a bit of good fortune.

The owl looks about ready to be released, but he has still to learn some of the finer points of aeronautics and tries to achieve impossible climbs. Every day he becomes more aggressive, clacking and blinking at us whenever he's approached. He should survive now if we continue to feed him.

THURSDAY 30TH

This morning could have been a morning in grey February—a cold wind and banks of threatening clouds sweeping in from the moor. And we're supposed to be haymaking.

The hens are laying much better—around the twenty-three mark a day from thirty-two hens. Several have gone broody and some seem more anxious to eat their eggs than to hatch them out. We've one Maran chick so far and there's a Maran gone broody under the oil tank on a nest of dried leaves. I have put six Maran eggs under her. They should hatch 21 days from now.

The vegetable garden is enjoying the "summer"—beans, potatoes, strawberries—it's a fine year for them. But there are very few apples on any tree, the frost took off the blossom, so the apple harvest won't take long this autumn.

Only one cow has been inseminated during this month:

Lucy—Thursday 2nd

HENS

When laying well, a good hen lays five or six eggs a week. This drops as the nights draw in during the winter. They begin to lay well again in February. They have a laying life of about 2 years, after which many become uneconomic to keep and have to be killed.
Whether laying or not they are fed daily with a mixture of corn and layers' pellets. Water and oyster shells are always available for them, the latter to harden the eggshells.

FOAL

Yesterday he was nowhere to be found
In the skies or under the skies.

Suddenly he's here—a warm heap
Of ashes and embers, wind-fondled.

A star crashed in from outer space—Blazed and burned out in one flash.
Now something is stirring in the remains. We call it a foal.

Still stunned,
He has no idea where he is.
His eyes, dew-dusky, explore gloom-walls and a glare doorspace.
Is this the world?
It baffles him, like a numbness.

He pulls himself together, getting used to the weight of things
And to that tall horse nudging him, and to this straw.

He's resting
From the first huge shock of light, the huge daze
Of the huge questions—
What has happened? What am I?

His ears keep on asking, gingerly.

But his legs are impatient,
Recovering from so long being nothing
They are restless with ideas, they start to try a few out,

Angling this way and that, feeling for leverage, learning
    fast—

And suddenly he's up

And stretching—as if a giant hand
Stroked him from nose to heel
Perfecting his outline, as he tightens
The knot of himself.
                    Now he comes teetering
Over the weird earth. His nose,
Downy and magnetic, draws him, incredulous,
Towards his mother. And the world is warm
And careful and gentle. Touch by touch
Everything fits him together.

And soon he'll be almost a horse.
Wanting only to be horse.
Each day pretending more and more horse

Till he's perfect horse. And unearthly horse
Surges him, weightless, a tossing of flame
Under sudden gusts,

And coils his eyeball and his heels
In a single terror—like the terror
Between lightning and thunderclap,

And curves his neck, like a sea-monster emerging,

And flings the new moons through his stormy banner
And full moons and dark moons.

Grazing on Wood Hill

JULY

FRIDAY 1ST

By lunchtime the sun was breaking the cloud and there was a warming breeze. We began baling on Great Eastern Hill in the late afternoon, leaving it as long as we could so the sun could dry it. It was not properly quailed, but it was not wet, and with the weather unpredictable we had to go ahead. By dusk and by the time the dew had come down, all except for the centre of the 12 acre field had been baled; and we would have done that also if the baler had not broken down and delayed us for 45 minutes. All the bales were stood up in stooks before we came home. In all there are nearly 2,000 bales on the field with another 200 or more still to bale, so again it is a very heavy crop.

The barley fields up on Berry are looking wonderfully well already. A warm spell during the next few weeks and we should have a good harvest there too.

BALING

The cut hay can only be baled when it has dried sufficiently. Hay that is stored damp is inclined to heat up. Some barn fires are caused that way.

The baler picks the hay up, compresses it, binds it with two cords and expels it as a bale of hay. Each bale weighs about 40 lb., but this is adjustable.

QUAILED

Grass dried into hay.

SATURDAY 2ND

John turned the remaining few swaths on Great Eastern Hill and had baled it all before mid-afternoon. It was perfect haymaking sunshine and we made the best of it. Front Meadow was turned and the outer swaths troned and baled before nightfall. Graham mowed the last hayfield, Essythorn; and then with Les Farley lending a hand we began loading in the bales from Berry. Five loads were brought back before we called it a day. Meanwhile the bales on Front Meadow needed standing up for the night.

In the middle of all this I fed the pigs and the lambs, and milked the herd with David. The evening milking interferes with haymaking, but once it was finished and the cows turned out we were loading down at Burrow again till dark.

The sun promises to stay, so with any luck we shall finish the hay harvest by mid-week.

STANDING BALES UP

Hay baled a little early needs to be dried through before storing. The bales are stood on end in fours leaning against each other. Thus the bales have some protection from rain and dew, the water running down the sides of the bales instead of soaking in. It is important to keep as much of the bale off the damp grass as possible.

SUNDAY 3RD

The priority now is to bring the hay inside. There were still eighteen trailer-loads to bring into the barn up on Berry, and more than a dozen loads still out on Front Meadow. The Dutch barn at Burrow is full to the rafters with Upper Redlands hay and today we began to fill up the Big Barn at

TRAILER-LOAD

Approximately 120 bales of hay in each load.

Parsonage. Every load takes two of us an hour—the loading on to the trailer, one with a pitchfork on the ground and one up there building the load; then the ride to the barn before off-loading on to the elevator and stacking tight and secure.

Towards evening after milking, Tulip began to look as if she might calve; her tail was up and she was away from the others in the field. Her waterbag broke as John drove her up into the side of the Big Barn. She couldn't seem to settle—the elevator was disturbing her, so we pulled the calf off—a Friesian heifer from one of our best milkers.

MONDAY 4TH

John turned Essythorn just as soon as the overnight dew had dried away, while the rest of us went on loading. With a cousin lending a hand, and Les Farley again this evening, we've managed to load almost all Front Meadow and about half the Great Eastern Hill up on Berry.

The hay from Front Meadow is filling up the Big Barn. Already Graham is stacking high up among the rafters with his head touching the roof. The elevator has chugged away efficiently all day, but the tyre on the hay turner burst and John had to stop to change it. Intensely hot all day. John is now worried the hay is drying up too quickly, so we have stacked the remaining bales in Front Meadow into small ricks so that the bales cannot dry out too much in the sun.

In the middle of all this Tulip upset things by developing milk fever. This morning she was lying down and refused to budge. So we called the vet. He gave her a calcium injection and said she should be on her feet and better by mid-day. She wasn't, so he was called out again. Another injection seemed to do the trick and she was on her feet within the hour and chewing the cud. The calf is fine.

It was warm outside until past 9 o'clock, so the baling stopped only for tea and milking. Hettie fed the pigs and saw to the sheep.

TUESDAY 5TH

I discovered a ewe in trouble this morning when I went to feed the sheep on Brinnen. She had pushed her head in through the bars of the creep—the adjustable bars are there so that only the lambs can poke their heads through to get at the feed. Her head was twisted round and she was stuck fast. With the greatest difficulty I loosened the bars and finally managed to wrench her head out. She was spluttering and wheezing and at first unsteady on her feet but she recovered.

Thunder came this evening while we were milking, and now the raindrops are ominously heavy. Up on Berry 1,300 bales have now been carried in and the remainder

ELEVATOR
An engine-driven elevator that can be set at any angle. The machine assists greatly in stacking and unloading bales or sacks. Some haymaking still has to be done by hand but the mower, the hay turner, the baler and the elevator have made it a lot easier and quicker.

CHEWING THE CUD
A cow has four stomachs. In the first stomach the chewed food is made up into balls of cud which are regurgitated into the mouth to be chewed again. The food is then broken down until it reaches the fourth stomach and that is where digestion happens.
A cow chewing the cud is usually in a healthy condition.

CREEP
A covered trough used for feeding.

stood in stooks. All the bales that were still left on Front Meadow are now safely in the Big Barn. Out on Essythorn, John turned again, troned and baled and we've taken in three trailer loads of that as well, and stood up the rest. If it does come on to rain hard, none of it will come to much harm. The hay tally so far stands at over 6,000 bales. We calculate we shall need 7,000 plus to winter the stock. We may just do it.

WEDNESDAY 6TH

We carried fourteen trailer-loads off Essythorn and stacked them in the Big Barn. There is some room left in there for the straw, but not much. It has been a bumper harvest, and the price of hay has fallen steeply as a result from over a pound a bale at the end of the winter to 50p a bale now. A warm evening, so loading went on until nearly 10 o'clock. Just the 1,000 or so bales on Berry now to carry in and it will all be done.

Out on back meadow the sheep have taken to the shade of the trees, moving round with the shadow whenever they feel the sun on their backs.

The flies are back with a vengeance—their favourite targets seem to be haymakers, sheep and cows.

THURSDAY 7TH

The last long day of baling. With a good harvest of straw in August there will hardly be a square inch of storage space left.

But we are tired of hay. The excitement of the first mowing and baling vanished some time ago when we knew it would be a good crop. From then on it became a long, hot, thirsty endurance test. Shearing the lambs will be a refreshing change.

Old Hyacinth was bulling again, but after four unsuccessful attempts at insemination, we won't be trying again.

LAMB'S WOOL

The wool off one lamb is worth around 75p, compared to £2.50 off an adult ewe. The fleece from a Suffolk ewe weighs between 5 and 7 lb., a lamb about 2 lb.

FRIDAY 8TH

The hay is done at last. We have brought in well over 7,000 bales from the four fields—and that's 1,000 more than we expected. So today we rested on our laurels. There were the pigs' pens to clean out—a task neglected during the haymaking, and David could take his time over the milking once again.

Graham discovered a ewe with Black Udder this morning when he was feeding the lambs. We have a case of Black Udder only rarely. Graham thinks he may have caught this one in time. He injected her with penicillin.

BLACK UDDER

A gangrenous infection in the udder of a ewe that will spread. The udder turns dark blue and the ewe will die if the condition is not treated.

Old Hyacinth

MAGGOTS
Blow-flies' eggs laid in the wool, hatch and turn into maggots. They must be removed by docking if the sheep are not to suffer. Sheep need a lot of care and attention if they are to remain healthy.

SATURDAY 9TH
It has turned overcast and heavy—good weather for maggots. If we had had the time we would have shorn the lambs before now, but everything had to wait until after the hay. Graham has picked out three with grey-looking behinds and twitchy tails. He drove them up into Lower Yard and docked them, covering the infested area with fly repellent. It must rank among the nastiest jobs on the farm, but from now on until they are shorn there may be two or three a day that will need treating.

The ewe with Black Udder has recovered completely.

I released the owl this evening after dark, leaving the shippen door open behind me. He is fully capable of sustained flight now, and can deal with a large dead mouse, pulling it to pieces in the manner born. Whether he can catch one is another matter, but we shall leave an evening meal on his perch each night until he no longer needs it.

SUNDAY 10TH
We picked out lambs that are fat enough for market. There is only room for about fifty in the cattle box, and we shall take those in tomorrow. The market for lambs slumped drastically a fortnight ago, but it is climbing back up. Selling the lambs now means fifty less to shear.

Everyone is recovering slowly after the hay harvest. The last trailer is standing still unloaded in the barn—no one feels much like dealing with it yet.

The owl meat was gone this morning, and tonight we again put more down.

MONDAY 11TH

Flushed with success at the fat market. The fifty-three lambs Graham took in made £22.50 each, and the four pigs about £35 on average. Multiplied that sounds a lot of money, but there is a water bill of £200 to pay, a feed bill of astronomic proportions and the mini pick-up will need replacing soon. The inflow of large sums of money provides some relief rather than a surge of excitement.

Lamb-shearing in the afternoon. The lambs are easier to subdue than the ewes, easier to catch and there is less wool to take off; the wool is not long enough to form a fleece so we just bag it loose.

TUESDAY 12TH

The dust has been settled by a heavy shower of rain around mid-day, but we had the lambs inside already by that time, so the wool was dry for shearing. Graham and David got through several by teatime.

I spent the afternoon weeding Lamb's Tongue from the kale field. The field had been sprayed against weed back in the spring, but Lamb's Tongue seems particularly resilient. The kale is growing well enough but weeding will give it more room as the plant grows.

LAMBS TONGUE
A weed that grows tall and fast. If allowed to grow it will crowd the crop.

WEDNESDAY 13TH

From 4th July to the 11th August has come to be known as the "dog days". Any cut or scratch on the sheep and it won't heal until after 11th August. It is the heavy heat and the flies, and the combination of the two aggravates any wound.

Too wet for any lamb shearing, so it was another day for weeding. I pulled more Lamb's Tongue out of the kale field, and David finished off cutting the weed down on the Marsh. Graham cut nettles up around the top of Back Meadow with the hook.

That last load of hay was finally unloaded—it was the earliest anyone could face it.

THURSDAY 14TH

Drying again all day and thirty more lambs shorn by this evening. A few had maggots, but we are keeping them under control.

A Jersey cross steer has gone lame down on Ferny Piece. It

could be Kibe, a foot-rot condition. Graham drove him up to the yard to keep an eye on him.

The weed seems to be growing like no other year—but then it always seems like that. The hedgerows along the lanes are far advanced, narrowing the lanes so that in places there is barely room for a car to pass without being brushed on either side by the hanging grasses. We don't cut hedges until August usually, but they need it now.

KIBE
Cattle are liable to foot-rot conditions particularly if the feet have been softened in the wet. If neglected the foot-rot could spread upwards into the joints. If possible lame cattle should be kept on hard ground during treatment.

FRIDAY 15TH
The day was spent lamb shearing. Quicker it may be than the ewes, but it's a fiddly, awkward job. For a start you have to bend lower to get a firm grasp on the lamb. We have only sixteen more to do now, and we should manage those tomorrow if the weather keeps fine.

Buttercup calved out on Front Meadow in the afternoon sun. She is one of our three Guernseys, and she had been inseminated with a Charolais bull. The result is a lively, bulky looking bull calf, a milk chocolate colour with Guernsey patches fore and aft. It is a useful cross—not up to Friesian or Hereford standards for beef—but then no Channel Island cross could ever be that. Buttercup is our quietest cow. Graham went up to her and stroked her with the new-born calf still wet from birth and she never paid him any attention.

GUERNSEY/CHAROLAIS CROSS
There are enough Channel Island cows in the herd, so a useful beef animal is the only practical alternative to a pure Guernsey heifer calf. Charolais provides the bulk to make a fair beef calf. Pure Channel Island cattle have little value when killed for beef. They are too thin.

SATURDAY 16TH
All the shearing is finished. One of the lambs had foot-rot and the foot had to be cleaned out and sprayed. We shall sell most of the lambs between now and the end of the summer except for the best ewe lambs that we will keep back for breeding. This year as well Graham has picked out the two best ram lambs which he will feed up and sell in the autumn as rams.

Petunia calved out on Front Meadow during the afternoon. Crossed with a Hereford, we were hoping for a big bull calf and that is what we got.

The lame steer was treated by the vet this morning for Kibe. He was given an antibiotic injection.

SUNDAY 17TH
Petunia and her calf were brought down to the dairy this morning. Her udder is fully stretched. The clusters are not tailored for it and find it difficult to hang on.

MONDAY 18TH
The owl was found lying dead by the gate at the bottom of Lower Redlands. He had eaten the food we put out for him.

We have no idea what happened to him. He died with his eyes wide open.

Wool grading at North Tawton. We loaded up the box to be in there by 1.30; 195 assorted ewe fleeces and several sacks of lambs' wool. In a fortnight we shall have the cheque from the Wool Marketing Board. At present all we have is a long list of gradings and prices per kilo.

There's a bullock that keeps straying across the brook and up on to Mr. Banbury's land. He has had to be brought back twice. It takes an age to pick him out and drive him home. But it always happens this time of year when the brook gets low, and when the flies get after the cattle.

John went off to help drill kale with Mr. Yelland on the next farm.

Dutch Elm disease has struck again this year. Within the last 2 weeks at least a dozen elms have turned yellow. They will all be gone before long, but there are young oaks and beeches already growing in their place.

WOOL GRADING

Wool is graded by the Wool Marketing Board according to the strength of its staple and according to its length. The wool should be clean and dry for grading.

Catching the sheep to check for maggots

TUESDAY 19TH

Hettie has reckoned that we should make £650 or thereabouts on the wool we took in to North Tawton.

Patricia—the cow that sucks her own teats—has calved again—a Friesian heifer calf. This is the beginning of concentrated calving over the next few weeks. They are mostly experienced cows so we should have no problems.

Already the Guernsey, Buttercup, has had to turn nurse cow, but we hope it's only temporary. She is quite difficult to milk at present with her udder stretched so wide. The cluster just will not stay on.

I saw a great ginger tomcat up on the owl's perch in the shippen this evening, sniffing around for meat. Clearly he was missing his daily meal. The owl must have died of starvation. Everything we thought he was eating must have been taken by the cat.

WEDNESDAY 20TH

Over the last 40 years John has lost roughly three-quarters of an acre to the River Torridge. Successive floods over a period of years have weakened the banks and eroded the Marsh. It has been a continual fight to keep the river out—concrete buttresses, great tree trunks and now stone. Graham, David and John took two loads—two boulders that is—and dropped them in place just out from the bank. A lot more will have to be done tomorrow before the bank is built up and secure.

Poppy calved on Front Meadow. She is a week early but that is nothing out of the ordinary. A bull calf.

RIVER BANKS

Water is an expensive commodity now, so that it is a great advantage for a farmer to have a river or a stream running through his land. On Parsonage there are ten fields with their own natural water supply. However, rivers seem to change course over the years and very often take valuable land with them.

THURSDAY 21ST

Poppy was not up on her feet this morning when David went in to bring her down to the dairy. Milk fever yet again. Like the others who have had it, she's a heavy milker. Mr. Hindson came out at once and injected her with calcium, and within a few hours she was on her feet and eating once again.

The Tamworth gilt farrowed at the same time, only three little pigs. It is hardly surprising; Tamworths are classified as a rare breed. They are a red-brown colour with a long pointed snout, but smaller and shorter than the Landrace. Clearly their productivity is not up to Landrace standards either, but then perhaps this farrowing is an exception. She is only a gilt after all—first time farrowing rarely produces a big litter.

Five Maram chicks hatched out this afternoon.

Opposite: *Strengthening the bank on the Torridge*

FRIDAY 22ND

More calving. Blossom was found this morning with a heifer calf up against the hedge on Front Meadow. Again she is early but all appears to be well. Blossom gives about 7 gallons a day when in milk—and that is just the sort we want a heifer from.

At last the niggling vacuum problem down in the parlour has been solved. The engineers came this evening and replaced the faulty switches. Now we can see again exactly how much each cow is giving before sending it up to the bulk tank. The milk yield is soaring again with the new influx of freshly calved cows—up above 180 gallons.

SATURDAY 23RD

Honey's calf was born early this morning. By the time we found her she was just walking with traces of damp still on her. Another heifer calf for the milking herd in $2\frac{1}{2}$ years. David is looking after seven calves now down at Burrow. Honey's was brought down for milking this evening. Two of the calves—the Charolais cross and the Hereford cross will go to market on Tuesday, but the rest are heifers and all of them look good enough to rear on at this stage.

The best sow is to farrow soon now. She has clearly been making her nest, so she will farrow during the night very probably. We need a good litter after the Tamworth's poor performance.

Colin Weeks, a forester, from the village came in to take down a scrub oak tree. John wanted some oak stakes to drive in around the stones down on the river bank. Colin cut out twenty stout ones and that should be sufficient.

SUNDAY 24TH

The sow had farrowed by the time I got to the pigs' house this morning. It is difficult to see how many there are with them all lying in a heap, but it looks as if there are eleven little pigs. The sow has plenty of milk and the little pigs look to be of even size. A good litter at last.

The barley out on Long Close does look a little thin, but up on Berry it is growing thick and high. With the weather cold and damp, it looks like a late corn harvest this year—maybe still another month away.

CORN

The change of colour from green to yellow shows that the corn is ripening and ready for harvesting.

MONDAY 25TH

We had Mr. Hindson in to look at Harriet. She has been moving even more slowly and awkwardly than ever. He pared her bad foot, but that was all that could be done. He does not hold out much hope for her.

Graham took in four of the older ewes that are no longer suitable for breeding. Two of them are wrong in the udder and the other two are broken mouth. They made £17 each. He took in three fat pigs and a gilt that seemed incapable of breeding—she had been to the boar several times. They made around £37 a head.

BROKEN MOUTH
A sheep that has lost most of its teeth.

Bonny lost her calf last year, so John had her in to keep a good watch on her. Hettie came out to help and between them they took off a heifer calf. It was quite a struggle though. Usually once the head is out the rest will come easily, but this one got stuck half-way out. Bonny is well and the calf is female, Friesian and fit.

TUESDAY 26TH

Caroline calved, but after producing a fine heifer, she fell down with milk fever and we had to have the vet in. She recovered within a couple of hours. The calf pens are filling up, in spite of the two that went to market this morning. The Hereford cross bull calf made £78, and the Charolais cross bull calf made £62. The Charolais cross was looking more and more bony and more like a Guernsey each day; but he must have looked promising enough because that is a good price for a Charolais cross Guernsey.

CALF PENS
They should be warm but well ventilated. The bedding is a build-up of dung and straw, but the front of the pens is cleaned out regularly.

David and John have picked more Lamb's Tongue than they care to mention off the kale field. Most of what is left is the broad-leafed kale or swede, but no one is going to look that closely any more. It is a back-breaking and tedious job.

Three more loads of rocks were taken down to the river.

WEDNESDAY 27TH

The runt among the sow's eleven little pigs died during the night. It had been losing ground every day, and no one expected it to live for long. The others are all well. Graham has deloused them and injected them. The sow has enough milk to feed all of them.

The heifer calves have been split up. Four are up at Parsonage with the nurse cow, Buttercup; and the younger four are still down at Burrow feeding from the bucket. We are milking forty-eight cows down there now and the milk yield seems to creep up daily.

Yet more weed-cutting on Burrow Brimclose—mostly docks and the dreaded thistle.

THURSDAY 28TH

The wool returns came this morning, showing that Hettie had slightly underestimated the figure. It came to nearly

£700, but then some of that goes back in V.A.T.

John has been considering the balance of the stock. He is thinking we should have a few more beef cattle and fewer breeding ewes. We have the hay now. We are thin on steers at present and it is a side of farming John has always enjoyed. He intends to sell about fifty of the ewes and Graham and Bounce went out in the afternoon to pick them out.

The weed in Lawn Field was cut this morning, and we will be letting the cows in there tomorrow to pick it all up. Once cut, thistles take on a sweetness and the cows very soon tidy up a field.

More stones hauled down to the river. We shall be at it for some time.

FRIDAY 29TH

Summer is back again and the flies with it.

Graham took Harriet into the abattoir in Hatherleigh this morning. The treatment to the foot had not really made any difference. At every milking she took more and more time coming up the slope into the parlour. Her milk yield had fallen drastically. We had doubts whether she would make it up into the box, but she did. It is always sad to see a cow go. She had been a good milker and very gentle to handle.

Bounce

While David and John were down at the river, Celandine, the kicking Guernsey, calved. Graham helped to take off a Charolais cross heifer calf. She will be worth a lot less than a bull, but Celandine has a wonderful bag on her and if we can tame her with that kicking bar, then we have a chance of making something of her.

SATURDAY 30TH

Graham picked out the last of the forty-eight ewes we shall be selling off next week, and it has not been easy because most of them are fine breeding ewes. With the heat back the lambs grazing on the Marsh were beginning to suffer from the flies down by the river. They had eaten most of the grass there anyway, so Graham brought them back up to Essythorn, where there is more shelter.

RIVER FLIES

Marsh is the lowest field, and in the warm weather at this time of year the warble flies are around troubling the stock.

Celandine is back down in the parlour and has already established that she has no great respect for the kicking bar. It was a struggle, but David milked her out successfully.

SUNDAY 31ST

The barley has turned yellow. Perhaps I only noticed it because the sun was shining strong on the fields for the first time for weeks. The yellow is at least a promise that it is ripening well. John feels it may be another 3 weeks still until we can harvest. It will need a lot of warm sunshine to hurry it along and make up the lost ground of the early summer cold.

The following cows have been inseminated during this month:

Ruby—Thursday 7th
Sorrel—Tuesday 12th
Tulip 2nd—Tuesday 12th
Hyacinth 2nd—Wednesday 13th

COMING DOWN THROUGH SOMERSET

I flash-glimpsed in the headlights—the high moment
Of driving through England—a killed badger
Sprawled with helpless legs. Yet again
Manoeuvred lane-ends, retracked, waited
Out of decency for headlights to die,
Lifted by one warm hind leg in the world-night
A slain badger. August dust-heat. Beautiful,
Beautiful, warm, secret beast. Bedded him
Passenger, bleeding from the nose. Brought him close
Into my life. Now he lies on the beam
Torn from a great building. Beam waiting two years
To be built into new building. Summer coat
Not worth skinning off him. His skeleton—for the future.
Fangs, handsome concealed. Flies, drumming,
Bejewel his transit. Heatwave ushers him hourly
Towards his underworlds. A grim day of flies
And sunbathing. Get rid of that badger.
A night of shrunk rivers, glowing pastures,
Sea-trout shouldering up through trickles. Then the sun
again
Waking like a torn-out eye. How strangely

He stays on into the dawn—how quiet
The dark bear-claws, the long frost-tipped guard hairs!
Get rid of that badger today.
And already the flies
More passionate, bringing their friends. I don't want
To bury and waste him. Or skin him (it is too late).
Or hack off his head and boil it
To liberate his masterpiece skull. I want him
To stay as he is. Sooty gloss-throated,
With his perfect face. Paws so tired,
Power-body relegated. I want him
To stop time. His strength staying, bulky,
Blocking time. His rankness, his bristling wildness,
His thrillingly painted face.
A badger on my moment of life.
Not years ago, like the others, but now.
I stand
Watching his stillness, like an iron nail
Driven, flush to the head,
Into a yew post. Something
Has to stay.

AUGUST

MONDAY 1ST

The heatwave is still with us and the water levels in the wells have fallen considerably. With the majority of the water-troughs up at Parsonage being supplied from wells, we have had to make more use of the brook. David and Graham moved the fences down on Watercress over to the other side of the brook so the cattle now have access.

The corn is ripening, the milking is profitable at around 180 gallons a day, the calves come out the right sex, the hay is in, and the sun is shining.

TUESDAY 2ND

Emma's last calf is fully weaned now, and she has been brought up to Parsonage, along with the two older calves from Burrow.

Tupping will begin shortly. This year we shall stagger it better, so that we shall have our lambing spread more evenly over January and February and March. There will be fewer ewes, so it should be a lot easier on everyone. We shall be looking around soon for a ram lamb to put in with the ewe lambs we are keeping for breeding.

RAMS
To avoid inbreeding a new ram has to be introduced each year if the ewe lambs are related.

More stone shifting down at the river. A few floods with the silt washing over and it will look like part of the bank.

WEDNESDAY 3RD

Several trips down to the river again, with stones, tree stumps and wire to bind it all in. John has the bit between his teeth now, and there will be no let up until the bank is made sound enough to hold the river. It is hard, hot work and the flies never leave you alone.

Graham went on weed-cutting with the finger mower and saw to the evening milking. Celandine seems to be getting used to the kicking bar. She has stopped struggling against it and now uses her tail to wreak her revenge on us.

THURSDAY 4TH

As expected the "dog days" have brought their toll of infection for the sheep. It has become a regular part of Graham's day now to check them for cuts and scratches and to treat them with antiseptic cream when necessary. There are about half a dozen to treat every day.

Some of the calves are scouring. It is the ones on the bucket that seem to be affected. Those sucking Buttercup seem in fine fettle. Graham has drenched them.

Mr. Holwill brought up his digger from the other side of the village to break up a pile of packed-in rubble and earth John wanted to take down to the river. Graham had to leave his weed-cutting and paring to join in the river works. All hands were needed.

FRIDAY 5TH

Poogly is causing David some anxiety. When she had her colic trouble some weeks ago, the vet said he thought she was not in calf. She hasn't yet come bulling again and David thinks there may have been a "silent" bulling, that is one which has gone unnoticed. He will have a test done on her soon to be sure one way or the other.

Graham picked out thirty-five of the best ewes from the flock this afternoon. They will be the first to go to the ram.

EWE LAMBS

Only the best ewe lambs are kept for breeding. The rest are sold fat. It is only by continuous selection that the quality of the flock can be improved.

SATURDAY 6TH

Tupping this morning. In driving rain we drove the first thirty-five ewes up to Berry with their ram. The sun shone as we set off, but a great grey cloud came down on us from Dolton Beacon and we were drenched by the time we got the sheep up Berry.

Then Charlie Weeks came with his son Colin, to help us with the tree trunks down at the river. The river has come up with the rain, but not enough to damage our barrier of wood and stones.

The Dutch Elm disease is going to be worse than we thought. Almost every week now I notice another one yellowing at the top. Burrow Lane is carpeted with the tell-tale yellow leaves. The tallest elm in Burrow Lane, maybe 100 feet, has just turned and there are at least five other big ones that will have to come down, as well as several smaller ones. It is a good thing we have plenty of oaks about.

SUNDAY 7TH

The two cows inseminated today both give cause for concern. Herma 2nd has had to be inseminated for the third time and Lucy for the fourth. There is always the fear with a heifer that she may be barren, and with a cow that something has gone wrong.

The corn is 3 feet up and a gentle yellow, the kale field a deep dark green and there is a glorious absence of mud.

MONDAY 8TH

Old Hyacinth went to market this morning. Her milk was lessening each day and the market is good for barren cows. She made £220, only £20 less than John paid for her and we have had a fine bull calf and a heifer out of her and a lot of milk. But Hyacinth was not any old cow. She was once the leader of the herd, and always easy to handle in the parlour. We will miss her.

Graham took in the old sow that had a poor litter last time. She made £61, but the fat pig he took in with her he brought back. No one offered enough. Five old fat sheep, no use for future breeding went for over £20 a piece.

Daisy, the last, and the only Guernsey that's really sound out of the original four, has calved. It is a Charolais cross bull calf, big and broad and light fawn in colour.

Graham and David castrated a litter of pigs and then went off down to the river with John.

TUESDAY 9TH

Hot sun by 9 o'clock this morning. Everyone else seems to be out cutting back their hedges and the lanes have widened by several feet. But the river is more pressing for us and everyone was down there most of the day, pulling in the brushwood and filling in.

Graham went to market in the morning just to take in Celandine's Charolais cross heifer. She made £36 and Graham is pleased. She is not a good looking calf, too much bone about her.

The calf born yesterday looks to be a good one and Daisy has a huge udder. They came down to the dairy this morning.

WEDNESDAY 10TH

Graham took a casualty lamb into the slaughter-house in Hatherleigh. He found it down by the stream unable to stand. Graham thought it best to take it in before it lost any more condition. This has been a good year for the sheep—we have lost very few and they have kept well.

When he got back Graham began drenching all the lambs again against stomach worm, and he was out in the afternoon making a start on the ewes as well.

THURSDAY 11TH

Sun and a deep blue sky with scattered mackerel clouds—a good sign. Graham went over to Berry to drench the rest of the ewes. They are in fine condition, with plenty of grass and sunshine, just as they should be for tupping. It should mean more lambs for each ewe and strong ones.

The cows are grazing again up on Front Meadow—the best grazing on the farm, but it is getting them to and from the milking parlour that is causing problems. They will not keep walking, but insist on stopping every few yards and rubbing. They are beginning to make it look untidy, pulling away the stones and rubbing bare patches down the lane. We have tried everything to stop them and to keep them moving. Graham goes in the middle of the herd flailing a branch, with David and myself behind and Bounce barking all round. But the cows pay little attention to any of us.

Mr. Holwill, our hedge-trimming contractor from Iddesleigh, began work on the hedges this morning. There are a few miles of road hedges to be cut and many more miles of field hedges. He does the cutting and we do the clearing up behind him.

CONTRACT WORK

With machinery so expensive and with limited labour there are some tasks on a farm that are often undertaken by a contractor. Hedge-trimming and combine harvesting are both done by Mr. Holwill.

Asleep by the woodpile

FRIDAY 12TH

Beth calved this morning in Back Meadow, but not before she had tried to join the rest of the herd, jumping a barbed wire fence on the way. She has torn one of her teats quite badly, so we shall have to be careful milking her over the next few days. It is a Friesian bull calf, and that is a pity because Beth is one of our heaviest milkers. Still Beth will be back with the herd tomorrow, bossing them all. In her absence the hierarchy has been a bit confused with no clear leader.

David and Graham went over to Berry in the morning to clear the hedge parings off the road. Mr. Holwill had cut about a mile of hedges and it took an hour with the buck rake to clear it all away.

HEDGE TRIMMER

Of two main types. One, like Mr. Holwill's, cuts the hedges cleanly but then the parings have to be picked up afterwards. The other is a flail hedge trimmer which cuts the parings into small pieces and forces them back into the hedge.

SATURDAY 13TH

Mr. Holwill finished trimming all the road hedges around Parsonage and made a start on the grass fields. The ground is still dust dry and so he began on the steep fields where it would be impossible to trim when the ground is wet.

Beth was a little touchy down in the dairy but David managed to treat the cut teat successfully, much against her will. The calf does not seem to understand that there is one teat he must not suck and gets kicked for his pains every time he goes near it. A slow learner.

ROAD HEDGES

The farmer is responsible in this part of Devon for cutting the hedges along the roads bordering his land. For the contractor there are several problems. It has to be done in high summer at the height of the tourist season when the lanes are busy. It often means stopping and manoeuvring every few minutes to let a car squeeze past in the narrow lanes.

SUNDAY 14TH

One of the fat Devon steers broke out into Berry Wood. We found the trampled fence, but it took a long time to find him. Berry Wood is very thick in places and he had wandered in amongst the forestry plantation. He was probably trying to escape the flies which are becoming unbearable for the animals in this heat.

The sheep too are suffering from them. We would like to dip them, but Government regulations state that they have to be dipped against Scab between September 1st and November 1st. If we do it before then we shall have to dip them again and we don't really want to dip them twice.

We are hoping for rain. The stream is very shallow and mucky, the cattle spend all their time in the cool in the water under the trees.

MONDAY 15TH

The corn on Little Eastern Hill could be ready for harvesting by the end of the week if the weather stays hot. We are hoping it will be dry enough—less than 16 per cent moisture—to have it blown straight into a bulk lorry. But it is a gamble. The lorry has to be ordered 2 or 3 days ahead, but if the weather then prevents combining on the day we've ordered it, the lorry still has to be paid for and that is £20 to £30—a lot of corn. We think we shall stick to bags—it takes more time but it's less risky.

Graham and Bounce brought the six steers back from Berry to Parsonage where there is more grazing. Bounce has matured into a fine dog. He kept the steers under control all the way back down the lanes. We can rely on him more and he seems to accept the responsibility well.

TUESDAY 16TH

John and David went down to the river to secure all the brushwood and parings with wire. Shortly after mid-day, while they were still down there, the rain came, and no ordinary rain but a heavy torrential downpour. John reckons they have done just about enough for it to hold if the river comes up.

Graham picked up the thorny hedge trimmings on Front Meadow and filled in the weaker places in the hedges with the best of it. The rest was burnt just before the rain.

In the market Daisy's Guernsey cross Charolais bull calf sold for £75, a huge figure for a calf that looked certainly more Guernsey than Charolais.

This rain is unlikely to affect the corn. The harvest will be postponed of course. If there were a heavy wind with the rain then it could be blown flat on to the ground and the stalks broken—flattened corn is difficult to combine. But today's rain came down straight and there is no hint of wind. All should be well.

WEDNESDAY 17TH

There were a few early windfalls from the apple trees and there seem to be a lot more apples than we first thought.

A pair of white pigeons are nesting in Old Barn in a hole in the old cob wall. These are special pigeons. They will be welcome, providing they do not multiply too fast.

The evenings are drawing in already.

THURSDAY 18TH

Graham's weed-cutting was interrupted by a difficult calving. A Friesian heifer, Dahlia, not due for a few more days yet, lay down up against the hedge and tried to calve. It was soon evident that the calving was not normal—there were two front feet protruding but no nose. Graham felt inside and found that the head was tucked down under. He managed to drive her into the shippen along with all the other heifers. He drove the others out and then after several unsuccessful attempts, chained her up while she was eating some cattle cake. Within minutes of lying down, the head came right and the rest slid out easily. A small Hereford cross Friesian heifer calf.

FRIDAY 19TH

Moving Dahlia and her calf from Front Meadow down to Burrow should have been easy enough. We had done it often before. But Dahlia turned out to be a wild heifer and she made it a major operation. They took all the wrong turnings they could. Milking was delayed as a result.

SATURDAY 20TH

The milk yield has fallen to around 160 gallons, probably because of the very wet weather. Beth has settled better and her sore teat is responding well to treatment. Patricia, the cow that sucks herself, has been supplied with a device that looks something like an enlarged bull's ring with teeth pointing outwards. Any attempt to suck herself is very painful so she doesn't do it any more.

David tried on a new rubber aprons for milking. It covers well but there is not much protection behind. You can't always be facing the right way when surrounded by eight cows.

SUNDAY 21ST

Just the milking, the pigs and the calves today. The farm looked after itself in between while everyone went to the sea for the day.

MONDAY 22ND

The weed-cutting machine failed us this morning. The connecting rod had been bent and the bearing was over-heating, so Graham took it into Mr. Vanstone in Monkokehampton who straightened it up for him. Graham was able to get on with cutting weed on Watercress in the afternoon.

David went back down to the river with John. The two of them spent the day building up the bank.

WEED-CUTTING MACHINE

At Parsonage a finger mower is used, a long pair of reciprocating saw blades.

TUESDAY 23RD

The lamb market was down this morning in Hatherleigh. Graham took in twenty ewe lambs but sold only ten at £26 each. He brought the other ten back. There was only one person interested and that is not much good at auction. There is the big annual sheep sale at South Molton tomorrow and many sheep farmers will be holding back until then. Beth's Friesian bull calf sold well enough though. He made £65.

We should be harvesting but the sky is heavy with rain again this evening and there is no sign of a change for the better.

Graham picked up a basket of mushrooms today on his way over to the sheep.

WEDNESDAY 24TH

It is most unlikely now that we shall have the bulk lorry for the grain with the weather as uncertain as this—there is no point in ordering it 2 days ahead. So when we do harvest it will be into sacks. John went off this morning to buy 300 new ones—they cost 18p each—we stayed under cover of Old Barn mending the used sacks with glue and patches. But at least we have enough now to be getting on with.

No one is looking for ideal harvesting weather now; one dry day and the combine harvester will be able to make a start. That is all we need.

CORN SACKS

They weigh 70 lb. when full. At least twenty sacks are needed for an acre of barley. The corn dries out well in sacks, and providing it can be harvested fairly dry there is no need for artificial drying. Artificial drying costs about £7 a ton.

THURSDAY 25TH

The barley is ripe and ready for harvesting. There is at least £3,000-worth of barley standing out there in the fields. Every day's delay now will prove costly. By this time last year the harvest was all done, the grain and straw inside under cover. Each patch of blue in the sky brings fresh hopes that it may be the beginning of a dry spell, but so far we have always been disappointed.

Graham is still having to treat several sheep that have maggots from time to time. Next year if the dipping regulations are the same we will dip in July as well as after the 1st September. The cost of dipping and the work involved is nothing compared to the condition the sheep lose and their suffering with the flies and maggots.

FRIDAY 26TH

There are only four really steep fields on the farm—Brinnen, Ferny Piece, Watercress and Wood Hill. The weed has to be cut on all of them. Graham has already cut Brinnen and today it was Watercress. He cut along the slope at an angle of about 30 degrees; any more and the tractor could have toppled over. It had to be done slowly and carefully, for once a tractor starts to turn over there's little the driver can do. For this we always use the big tractor with the safety cab. It took Graham most of the day to finish Watercress; and there are still a few yards near the hedge at the top where it's too steep to cut except by hand.

David spring-cleaned the dairy and began repainting the walls where the paint has flaked badly.

SATURDAY 27TH

It is just 3 weeks since tupping began, so it was time for a change over. The thirty-three best ewes, due to lamb now in

BRA

January, went in with the second ram, and we brought back the best ram to go in with the next fifty. In another 3 weeks we shall introduce the last fifty ewes and then finally the twenty-five ewe lambs we shall be keeping on. We shall be lambing next year well into April.

That last litter of pigs has turned out to be almost flawless. All of them have the long Landrace back, and full hams. We shall certainly keep three or four of the gilts back for breeding. The little Tamworths look dark and shorter alongside them, although they are growing away well enough.

A day of violent thunderstorms after a glorious clear early morning. The Produce Show in the village was crowded in spite of it; and the jams, cakes, fruit and vegetables filled the tables in the village hall.

Meanwhile, back on the farm there was a power cut and David had problems linking up with the reserve power generator for the parlour. He had just fixed it up when the power came back on again.

SUNDAY 28TH

Clear skies again and we can think about the harvest. Another day like this and the corn will be dry enough. John rang up Mr. Holwill, the contractor, and he said he could be ready to make a start by tomorrow afternoon. The forecast is fair, so now at last we may be in with a chance.

The milk yield has soared after all this rain. The grass was looking a little thin and shrivelled 2 weeks ago, but now the lushness of early summer is back. David has increased their cake ration as well to compensate for the lack of grass. He recorded 185 gallons this evening.

Dahlia's calf has the scour, and it is quite bad. It is a struggle to get the calf to accept the bottle of green scouring drench—only a proportion went down this morning. David gave her the second dose after milking this evening, so she should recover quickly now.

COMBINE HARVESTER

A machine that cuts, and threshes the corn. At a cost of more than £10,000, many farmers prefer to hire a contractor for this work. It costs about £8 an acre.

MONDAY 29TH

Harvesting under grey skies. The later corn at Parsonage was too wet to harvest, the corn still soft from the rain, but up on Berry where it was drilled earlier, and where the wind had dried it more quickly the barley was ready to harvest. In some places the corn was overripe and bent over so that it almost touched the ground. Any further delay and the corn would have been on the ground and irretrievable.

Mr. Holwill's combine harvester began work on Little Eastern Hill at 3 o'clock. By 5.30 that field was finished and there were still 2 hours of light left. So he started on the Dutch Barn Field. It was then that things started to go

Opposite:
Loading the barley onto the elevator with Mr. Farley

wrong. First the combine broke down—something stuck in the elevating system, and then a tyre on the baler exploded with a bang and put that out of action for the rest of the evening. There was just time to load up the straw that John had managed to bale, before the sun went down.

The combine harvester does everything except bale the straw. Chunking rhythmically like a great green paddle

steamer it cuts the corn, removes the grain and disgorges the straw. Once the tank is full, Graham pulls up alongside the moving combine and the corn is funnelled off into the grain trailer, then from the trailer into the jute sacks.

By nightfall we had over 180 bags filled—most of them under cover in the barn up there and the rest covered by a tarpaulin in case it rains overnight.

Making a load

TUESDAY 30TH

There had been rain overnight, so we had to wait all morning for the corn to dry out before we could begin combining Dutch Barn Field. The corn was thick and had ripened well, but everywhere it was bent low, in places broken off and lying down.

The corn itself was well dried and like yesterday's there will be no need to have it dried artificially. After 3 days in the sacks it will dry off of its own accord. There was a certain amount of dust in it and although the straw is not as bright as we would like it, at least it's another field done and a fair harvest at that; 200 sacks off the field and a lot of straw.

Walking around the field afterwards you can see there is some barley lying in the soft earth. But it will not be wasted. We will bring the pigs up here later on and they'll be able to live off the barley we leave behind for some weeks.

The calf with scour has improved, and the milk yield continues to rise slowly.

Pigs out on back meadow

Harvest Tea

WEDNESDAY 31ST

The first eighty sacks of barley have been taken away to the corn merchant at Dolton. The rest of it is still out in the fields again tonight along with much of the straw. Today was to have been spent tidying up the two fields we had already harvested, and in particular bringing in the hundreds of straw bales. There was sunshine first thing but by the time we got up to Berry the grey clouds were advancing again from the moor. We did manage to get the barley away and to build most of the straw into temporary ricks, but then the rain came sweeping in and there was no more we could do. We brought back a load of damp straw.

There has been some sheep worrying going on in the parish. A farmer had three ewes killed only last weekend, and today Elizabeth saw two black and white dogs chasing our sheep on Front Meadow. They had split the flock by the time she chased them off. No damage done, but if we catch them at it again they will have to be shot.

We have still three fields to harvest.

SHEEP WORRYING

A good ewe is worth over £40. A dog can kill several once he gets in among the flock.

The following cows have been inseminated during this month:

Herma 2nd—Sunday 7th

Lucy—Sunday 7th

Lucille—Tuesday 9th

Clover—Tuesday 9th

POSTSCRIPT

The farming year never fits snugly into 365 days. This year both the sheep dipping and the corn harvest were late, but by mid-September everything was done.

It was on the 1st September that we dipped the sheep. The Ministry have to be informed 5 days before so that they can inspect if they wish. The entire flock, ewes, rams and lambs were driven down Burrow Lane to Mr. Yelland's dipping pit beyond Burrow. It took a morning to dip them all. Each one

Coming in for sheep-dipping

has to be immersed in the pit for a minute and during that time must be submerged so that every part of the sheep is treated with chemical that prevents Scab and Fly strike.

In the days following we harvested in cold, temperamental weather, never sure how long the rain would hold off; but it was a good harvest. From the 35 acres of corn, we harvested over 54 tons of barley and nearly 2,000 bales of straw. The quality of the straw was not as good as we had hoped owing to the belated harvest; but it will make good bedding, though there will not be much eating straw from it.

Back at Parsonage after the sheep-dipping

John is pleased with the grain harvest. Two of the fields were on their third year of corn and one had been under the plough for 5 years. We do not expect a very high yield from fields after 2 years of ploughing. In spite of the rain there was very little we could not pick up and the corn dried out satisfactorily in the sacks without any need for artificial drying. We shall keep back 3 tons for next year's seed. Most will be sold for around £65 a ton, and some kept back for cattle and pig feed.

There is no beginning and no end to a farming year. For John, Hettie, Graham and David it is an unending cycle. Already the erishes are being ploughed up, Long Close is drilled with rye grass for the cows next spring, the apples are being picked and the leaves on the trees beginning to turn. We expect the first frost any night. And so it will go on, all around the year.

Opposite:
John dipping sheep with Mr. Yelland

INDEX OF NOTES

A.I. 37
Animal feeding stuffs 34

Baling 173
Banks 106
Barns 65
Barren cow 123
Beef calves 102
Blacksmith 73
Black Udder 175
Blindness 74
Bloat 35
Blowing cake 91
Brucellosis 123
Broken mouth 183
Buck rake 23
Bulk tank 98
Bullocks 58
Bushel 127
Butterflies 107

Cab away 91
Cake 24
Calf pens 183
Calving 38
Castrating pigs 67
Cattle behaviour 56
Chalk 34
Charolais 37
Chewing the cud 174
Circular saw 23
Clock change 39
Club Day 147
Cluster 42
Colic 146
Colostrum 27
Combine drill 124
Combine harvester 197
Combs and cutters 152
Concentrates 31
Concrete 91
Contract work 191
Corn 182
Corn sacks 195
Couple 130
Creep 174
Cross breeding 76
Crush 56
Cuckoo 144
Cultivating and drilling 129

Dairy 32
Dehorning 42
Delousing 159
Devon 37
Disc harrows 116
Docking 135
Drags 116
Drilling 124
Dry cows 24, 161
Dung spreader 29
Dutch Elm Disease 136

Elevator 174
Emergency bulk tank 103
Erish 52
Ewe hoggets 113
Ewe lambs 189

Farm sale 15
Fencing 118
Fencing and ditching 21
Fertilizer 25
Fluke and worms 151
Foot-rot 27
Friesian 37

Gilt 58
Grading 68
Grass 26
Guernsey 37
Guernsey/Charolais cross 178

Haflinger 32
Harrowing 126
Hay 21, 144, 165
Hay barns 69
Headlands 19
Hedging 29
Hedge trimmer 192
Heifer 16
Hen breeds 55
Hens 169
Hereford 37
Humane killer 149
Hygiene 53

Iddesleigh 147

Jersey 37

Kibe 178
Kicking bar 157

Lactation 22
Lamb doubles 89
Lamb drenching 27
Lamb dysentery 97
Lambing 72, 80, 90
Lamb pellets 160
Lambs Tongue 177
Lambs' wool 175
Landrace 58
Leader 49
Leather-jackets 148
Leaving the hay 131
Ley field 116
Link box 87

Maggots 176
Market 87
Mastitis 31
Metric 31
Milk fever 50
Milk prices 144
Milk recording 66
Milk yields 70
Milking parlour 22
Mixed farming 70
Mower 161
Mowing grass 159

Nitrogen 133
Nominated bulls 38

Owl 163

Paperwork in the market 144
Pig breeds 58
Pig meal 52
Pigs 41, 73
Ploughing 15, 116
Pneumonia 103
Point of lay pullets 25
Poisoning 142
Price of cows 74
Prolapse 130

Quailed 173
Quarter 93

Rabbits 151
Rams 83, 188
Return 123
Rhode Island Red 55
Ringing lambs 113
Ringworm 35
River banks 181
River flies 185
Road hedges 192
Rolling 132
Root crop 116
Runt 77

Scour 33
Scraper 87
Separation 57
Sheep breeds 21
Sheep feeding in winter 66
Sheep grazing 103
Sheep Scab 51
Sheep worrying 201
Shippen 38
Shows 142
Sick animals 71
Slurry 80
Slurry pit 53
Spraying 153
Staggers 146
Standing bales up 173
Steeping 27
Steers 99
Stone bank 160
Store market 26
Strip 26
Stripping 31
Subsidy 89
Suffolk sheep 90
Swath 163

Tamworth 58
Tilling 116
Tractor fuel 42
Tractors 27
Tractor tyres 162
Trailer-load 173
Trees 152
T.T. testing 97
Tupping 15
Turning, troning and baling 166
Two-tooth 95

Udder washing 66
Undersowing 149

Vet, the 16

Warbling 55
Weaners 26
Weed cutting 158
Weed cutting machine 194
Wells 58
Wild flowers 149
Winnowing machine 127
Winter feeding 57
Wooden Tongue 117
Wool grading 179
Wrench 112